Chemistry, Air, and Climate

A ChemCom Module

American Chemical Society

Kendall/Hunt
Publishing Company
Dubuque, Iowa

ChemCom Personnel

Principal Investigator:
W. T. Lippincott
Project Manager:
Sylvia Ware
Chief Editor:
Henry Heikkinen
Contributing Editor:
Mary Castellion
Assistant to Contributing Editor:
Arnold Diamond
Editor of Teacher's Guide:
Thomas O'Brien
Revision Team:
Diane Bunce, Gregory Crosby, David Holzman, Thomas O'Brien, Joan Senyk, Thomas Wysocki
Editorial Advisory Board:
Glenn Crosby, James DeRose, Dwaine Eubanks, W. T. Lippincott (ex officio), Lucy McCorkle, Jeanne Vaughn, Sylvia Ware (ex officio)
Writing Team:
Rosa Balaco, James Banks, Joan Beardsley, William Bleam, Kenneth Brody, Ronald Brown, Diane Bunce, Becky Chambers, Alan DeGennaro, Patricia Eckfeldt, Dwaine Eubanks (dir.), Henry Heikkinen (dir.), Bruce Jarvis (dir.), Dan Kallus, Jerry Kent, Grace McGuffie, David Newton (dir.), Thomas O'Brien, Andrew Pogan, David Robson, Amado Sandoval, Joseph Schmuckler (dir.), Richard Shelly, Patricia Smith, Tamar Susskind, Joseph Travello, Thomas Warren, Robert Wistort, Thomas Wysocki
Steering Committee:
Alan Cairncross, William Cook, Derek Davenport, James DeRose, Anna Harrison (ch.), W. T. Lippincott (ex officio), Lucy McCorkle, Donald McCurdy, William Mooney, Moses Passer, Martha Sager, Glenn Seaborg, John Truxall, Jeanne Vaughn
Evaluation Team:
Ronald Anderson, Matthew Bruce, Frank Sutman (dir.)
Field Test Coordinator:
Sylvia Ware
Field Test Workshops:
Dwaine Eubanks
Field Test Directors:
Keith Berry, Fitzgerald Bramwell, Mamie Moy, William Nevill, Michael Pavelich, Lucy Pryde, Conrad Stanitski
Social Science Consultants:
Ross Eshelman, Judith Gillespie
Safety Consultant:
Stanley Pine
Art:
Rabina Fisher, Pat Hoetmer, Alan Kahan (dir.), Kelly Richard, Sharon Wolfgang
Copy Editor:
Martha Polkey
Administrative Assistant:
Carolyn Avery

This material is based upon work supported by the National Science Foundation under Grant No. SED-88115424 and Grant No. MDR-8470104. Any opinions, findings, and conclusions or recommendations expressed in this publication are those of the authors and do not necessarily reflect the views of the National Science Foundation. Any mention of trade names does not imply endorsement by the National Science Foundation.

Copyright © 1989 by American Chemical Society

ISBN 0-8403-5102-X

All rights reserved. No part of this publication may be reproduced, stored in a retrieval system, or transmitted, in any form or by any means, electronic, mechanical, photocopying, recording, or otherwise, without the prior written permission of the copyright owner.

Printed in the United States of America
10 9 8 7 6 5 4 3 2 1

CONTENTS

Safety in the Laboratory   iv

Chemistry, Air and Climate   330

A. Life in a Sea of Air   333
B. Investigating the Atmosphere   340
C. Atmosphere and Climate   361
D. Human Impact on the Air We Breathe   373
E. Putting It All Together: Is Air a Free Resource?   394

SAFETY IN THE LABORATORY

In *ChemCom* you frequently will perform laboratory activities. While no human activity is completely risk free, if you use common sense and a bit of chemical sense, you will encounter no problems. Chemical sense is an extension of common sense. Sensible laboratory conduct won't happen by memorizing a list of rules, any more than a perfect score on a written driver's test ensures an excellent driving record. The true driver's test of chemical sense is your actual conduct in the laboratory.

The following safety pointers apply to all laboratory activities. For your personal safety and that of your classmates, make following these guidelines second nature in the laboratory. Your teacher will point out any special safety guidelines that apply to each activity.

If you understand the reasons behind them, these safety rules will be easy to remember and to follow. So, for each listed safety guideline:

- Identify a similar rule or precaution that applies in everyday life—for example in cooking, repairing or driving a car, or playing a sport.
- Briefly describe possible harmful results if the rule is not followed.

Rules of Laboratory Conduct

1. Perform laboratory work only when your teacher is present. Unauthorized or unsupervised laboratory experimenting is not allowed.
2. Your concern for safety should begin even before the first activity. Always read and think about each laboratory assignment before starting.
3. Know the location and use of all safety equipment in your laboratory. These should include the safety shower, eye wash, first-aid kit, fire extinguisher, and blanket.
4. Wear a laboratory coat or apron and protective glasses or goggles for all laboratory work. Wear shoes (rather than sandals) and tie back loose hair.
5. Clear your benchtop of all unnecessary material such as books and clothing before starting your work.
6. Check chemical labels twice to make sure you have the correct substance. Some chemical formulas and names may differ by only a letter or a number.

7. You may be asked to transfer some laboratory chemicals from a common bottle or jar to your own test tube or beaker. Do not return any excess material to its original container unless authorized by your teacher.
8. Avoid unnecessary movement and talk in the laboratory.
9. Never taste laboratory materials. Gum, food, or drinks should not be brought into the laboratory. If you are instructed to smell something, do so by fanning some of the vapor toward your nose. Do not place your nose near the opening of the container. Your teacher will show you the correct technique.
10. Never look directly down into a test tube; view the contents from the side. Never point the open end of a test tube toward yourself or your neighbor.
11. Any laboratory accident, however small, should be reported immediately to your teacher.
12. In case of a chemical spill on your skin or clothing rinse the affected area with plenty of water. If the eyes are affected water-washing must begin immediately and continue for 10 to 15 minutes or until professional assistance is obtained.
13. Minor skin burns should be placed under cold, running water.
14. When discarding used chemicals, carefully follow the instructions provided.
15. Return equipment, chemicals, aprons, and protective glasses to their designated locations.
16. Before leaving the laboratory, ensure that gas lines and water faucets are shut off.
17. If in doubt, ask!

Chemistry, Air, and Climate

Contents

Introduction — 332

A. Life in a Sea of Air — 333
1. *You Decide:* Air, the Mysterious Fluid in Which We Live
2. Laboratory Demonstration: Gases—Much Ado about Nothing?
3. Air: The Breath of Life
4. *You Decide:* Air—Just Another Resource?

B. Investigating the Atmosphere — 340
1. Laboratory Activity: Chemical Composition of the Atmosphere
2. A Closer Look at the Atmosphere
3. *You Decide:* How Do Atmospheric Properties Vary with Altitude?
4. Air Pressure
5. Boyle's Law: How to Put the Squeeze on a Gas
6. Laboratory Activity: Temperature-Volume Relationship for Gases
7. *You Decide:* A New Temperature Scale
8. In Ideal Cases, Nature Behaves Simply

C. Atmosphere and Climate — 361
1. The Sunshine Story
2. The Earth's Energy Balance
3. Thermal Properties at the Earth's Surface
4. Drastic Changes on the Face of the Earth?
5. *You Decide:* Examining Trends in CO_2 Levels
6. Laboratory Activity: Measuring CO_2 Levels
7. *You Decide:* Reversing the Trend
8. Off in the Ozone

D. Human Impact on the Air We Breathe — 373
1. To Exist Is to Pollute
2. *You Decide:* What Is Air Pollution? Where Does It Come from?
3. *You Decide:* Identifying Major Pollutants
4. Smog: Spoiler of Our Cities
5. Pollution Control
6. Controlling Industrial Emission of Particulates
7. Laboratory Demonstration: How Industry Cleans Its Air
8. Photochemical Smog
9. *You Decide:* Automobile Contributions to Smog
10. Controlling Automobile Emission of Pollutants
11. Acid Rain
12. Laboratory Activity: Acid Rain
13. Acids and Bases: Structure Determines Function
14. Acid and Base Strength
15. pH

E. Putting It All Together: Is Air a Free Resource? — 394
1. Air Pollution Control: A Chemical Success Story?
2. Paying the Price
3. Looking Back and Looking Ahead

INTRODUCTION

Although we live *on* the Earth's surface, we live *in* its atmosphere. Air surrounds us just like water surrounds aquatic life. And like the Earth's crust and its bodies of water, the atmosphere is both a mine for obtaining chemical resources and a sink for discarding waste products. We use specific gases from the atmosphere when we breathe, burn fuels, and carry out various industrial processes. Humans, other living organisms, and natural events also add gases, liquid droplets, and solid particles to the atmosphere. These added materials may have no effect, or they may disrupt the environment locally—or globally.

Since human activities can sometimes lower the quality of air, the problem of air pollution leads to important social questions: Should air be considered a free resource? How clean should air be? How much will it cost to maintain clean air? What financial and environmental costs are associated with polluted air? Who should be responsible for pollution control? Citizens will continue to debate these and other questions at local, state, and national levels in the years ahead.

Acceptable answers to such difficult questions depend at least in part on understanding the basic chemistry of the atmosphere and the gases in it. We need to know about the atmosphere's composition and structure, general properties of gases, processes influencing climate, and natural recycling, which renews the atmosphere.

This unit explores the basic chemistry related to these topics, and examines current pollution control efforts. It offers another opportunity to sharpen your chemical knowledge and your decision-making skills. Take a deep breath and exhale slowly. The air you just moved so easily in and out of your lungs is our topic of study.

National Center for Air Pollution Control

This photo shows the Cincinnati City Hall (1963) half-way through a cleaning process to remove 34 years of dirt.

A LIFE IN A SEA OF AIR

It's said that we don't appreciate the value of things until we have to do without them or pay for them. This is certainly true of the Earth's atmosphere. Unless we are astronauts, long-distance runners, patients with diseased lungs, or scuba divers, we seldom think about the sea of colorless, odorless gases surrounding us. You are probably no more aware of the 14 kg or so of air you breathe daily than a fish is of the water passing over its gills. As long as air is present and not too polluted, we tend to take it for granted. We shall see that this has not proved wise.

How much do you know about the atmosphere and problems involving air quality? Test your knowledge by completing the following exercise.

A.1 *You Decide:* Air, the Mysterious Fluid in Which We Live

On a separate sheet of paper indicate whether you believe each numbered statement below is true (T) or false (F), or whether you are too unfamiliar with it to judge (U). Then, for each statement you believe is true, write a sentence describing a practical consequence or application of the fact. Reword each false statement to make it true.

Don't worry about your score. You won't be graded on this exercise; its purpose is to start you thinking about the mixture of gases in which you live.

1. You could live nearly a month without food, and a few days without water, but deprived of air, you would survive for only a few minutes.
2. The volume of a given sample of air (or any gas) depends on its pressure and temperature.
3. Air and other gases are weightless.
4. The atmosphere exerts a force of nearly 15 pounds on each square inch of your body.
5. The composition of the atmosphere varies widely at different locations on Earth.
6. The atmosphere acts as a filter to prevent harmful radiation from reaching the Earth's surface.
7. In the lower atmosphere, air temperature usually increases as altitude increases.
8. Minor components of air such as water and carbon dioxide play major roles in the atmosphere.
9. Two of the top 10 industrial chemicals are "mined" from the atmosphere.
10. Ozone is a pollutant in the lower atmosphere, but an essential component of the upper atmosphere.
11. Clean, unpolluted air is a pure substance.
12. Air pollution first occurred during the Industrial Revolution.
13. No human deaths have ever been directly attributed to air pollution.

14. Natural events, such as volcanic eruptions and forest fires, can lead to significant air pollution.
15. Destruction of materials and crops by air pollution represents a significant economic loss for our nation.
16. Industrial activity is the main source of air pollution.
17. The "greenhouse effect" is a natural warming effect that may become harmful through excessive burning of fuels.
18. In recent years, rain in industrialized nations has become less acidic.
19. Most air pollution caused by human activity originates with combustion.
20. Pollution control has not improved overall air quality.

When you are finished, your teacher will give you answers to these items, but won't elaborate on them. However, each item will be discussed at some point in this unit. The next section includes some demonstrations and at-home activities that illustrate the nature of the air you breathe.

A.2 Laboratory Demonstration: Gases— Much Ado about Nothing?

Since the gases in the atmosphere are colorless, odorless, and tasteless, you might be tempted to think they are "nothing." But gases, like solids and liquids, have definite physical and chemical properties. You and your classmates will observe several demonstrations that illustrate some of these properties. The twelve demonstrations answer four major questions about air.

- Is air really matter?   Demonstrations 1–3
- What is air pressure all about?   Demonstrations 4–8
- Why does air sometimes carry odors?   Demonstrations 9–10
- Is air heavy? Will it burn?   Demonstrations 11–12

To get the most out of the demonstrations, follow these steps:

1. Try to predict the outcome of the demonstration.
2. Carefully observe the demonstration and, if necessary, account for differences between your prediction and the actual outcome.
3. Try to identify the specific property of gases being demonstrated. Try to decide why gases exhibit this property.
4. Think of practical consequences on Earth that result from each property as exhibited by atmospheric gases.

Try the following additional activities at home, and record both the results and your explanation of them in your notebook.

1. Place two straws in your mouth. Place the free end of one in a glass of water and the other outside the glass. Try to drink the water through the straw. Account for your observations.
2. In the screw-on lid of a jar, punch a hole into which a straw will fit. Insert a straw and seal the connection with clay, putty, or wax. Fill the jar to the brim with water. Screw the top on and try to drink the water through the straw. Account for your observations. Explain how it is possible to drink through a straw.

The properties of the elusive stuff called air make it vitally important in our lives. Let's examine one facet of air's importance.

A.3 Air: The Breath of Life

Seen from the moon's surface, Earth's atmosphere blends with its waters and land masses, presenting a picture of considerable beauty. Other planets possess exotic beauty as well, but as far as unmanned explorations have determined, their beauty is not accompanied by support of life. Earth supports millions of species of living organisms, from one-celled amoebas to redwood trees and elephants.

In the first unit, on water, you considered the key role of water in supporting life on Earth. Although the present atmosphere was formed much later than the waters of the Earth, it also helps sustain plants and animals.

A major role of the atmosphere is to provide oxygen gas needed for respiration. The following activity will help you understand that role.

YOUR TURN

VI.1 Breath Composition and Glucose Burning

Below are questions about oxygen in inhaled and exhaled air, and in burning glucose in the body. All can be answered by using concepts you have already studied (moles and the molar relationships shown by chemical equations). To illustrate, consider how the following questions are answered.

In a certain sample of air, 0.80% of the molecules are those of carbon dioxide (CO_2). Assume there are 125×10^{20} molecules in each breath you take. How many carbon dioxide molecules pass into your lungs with each breath of this air?

It is convenient to express percent as number of molecules per 100 molecules. The air sample is 0.80% CO_2, so there would be 0.80 CO_2 molecule per 100 air molecules. A breath contains

$$\left(\frac{125 \times 10^{20} \text{ air molecules}}{1 \text{ breath}}\right) \times \left(\frac{0.80 \text{ } CO_2 \text{ molecule}}{100 \text{ air molecules}}\right)$$

$$= 1.0 \times 10^{20} \text{ } CO_2 \text{ molecules/breath}$$

How many moles of carbon dioxide are present in each breath? How many grams? There are 6.02×10^{23} molecules per mole of any substance and the molar mass of CO_2 is 44.0 g/mol.

First, we can find the moles of CO_2 from the molecules of CO_2 in each breath.

$$\left(\frac{1.0 \times 10^{20} \text{ } CO_2 \text{ molecules}}{1 \text{ breath}}\right) \times \left(\frac{1 \text{ mol } CO_2}{6.02 \times 10^{23} \text{ } CO_2 \text{ molecules}}\right)$$

$$= 1.7 \times 10^{-4} \text{ mol } CO_2/\text{breath}$$

Then, we can find the quantity in grams:

$$\left(\frac{1.7 \times 10^{-4} \text{ mol } CO_2}{1 \text{ breath}}\right) \times \left(\frac{44.0 \text{ g } CO_2}{1 \text{ mol } CO_2}\right) = 7.5 \times 10^{-3} \text{ g } CO_2/\text{breath}$$

Each breath contains 7.5×10^{-3} g, or 7.5 mg of carbon dioxide.

The following exercise will help you solve problems such as how much oxygen or glucose you need each day, each hour, or each minute.

How many grams of oxygen will be consumed in completely burning 650 g of methane (CH_4), the principal component of natural gas? The equation is

$$CH_4(g) + 2\ O_2(g) \rightarrow CO_2(g) + 2\ H_2O(g)$$

Mole and molar mass were introduced in Unit Two, Part C.3, page 111.

The equation shows relative numbers of moles. Therefore, the mass given must be converted into moles. Taking 16.0 g/mol as the molar mass of methane gives

$$650 \text{ g CH}_4 \times \left(\frac{1 \text{ mol CH}_4}{16.0 \text{ g CH}_4}\right) = 40.6 \text{ mol CH}_4$$

The chemical equation shows that for every mole of methane that burns, two moles of oxygen gas are required, so we need 2 × 40.6 mol of O_2, or

$$40.6 \text{ mol CH}_4 \times \left(\frac{2 \text{ mol O}_2}{1 \text{ mol CH}_4}\right) = 81.2 \text{ mol O}_2$$

which is equivalent to

$$81.2 \text{ mol O}_2 \times \left(\frac{32.0 \text{ g O}_2}{1 \text{ mol O}_2}\right) = 2.60 \times 10^3 \text{ g O}_2$$

In burning 650 g of methane, 2.60×10^3 g of oxygen will be consumed. This answer could have been found with a single math setup as follows:

$$650 \text{ g CH}_4 \times \left(\frac{1 \text{ mol CH}_4}{16.0 \text{ g CH}_4}\right) \times \left(\frac{2 \text{ mol O}_2}{1 \text{ mol CH}_4}\right) \times \left(\frac{32.0 \text{ g O}_2}{1 \text{ mol O}_2}\right)$$
$$= 2.60 \times 10^3 \text{ g O}_2$$

Burning 1 g of methane releases 54 kJ of heat energy. How much heat energy is released by burning 650 g of methane?

$$650 \text{ g CH}_4 \times \left(\frac{54 \text{ kJ}}{1 \text{ g CH}_4}\right) = 3.5 \times 10^4 \text{ kJ}$$

Burning 650 g of methane releases 3.5×10^4 kJ of heat energy.

1. Consider Figure VI.1 which compares the composition of the gases we breathe in with that of the gases we breathe out.
    a. Summarize the changes in the composition of air that result from its entering and then being exhaled from the lungs. How do you account for these changes?

Compare the energy from burning 650 g of CH_4 with that from burning a similar mass of glucose. See Question 2c.

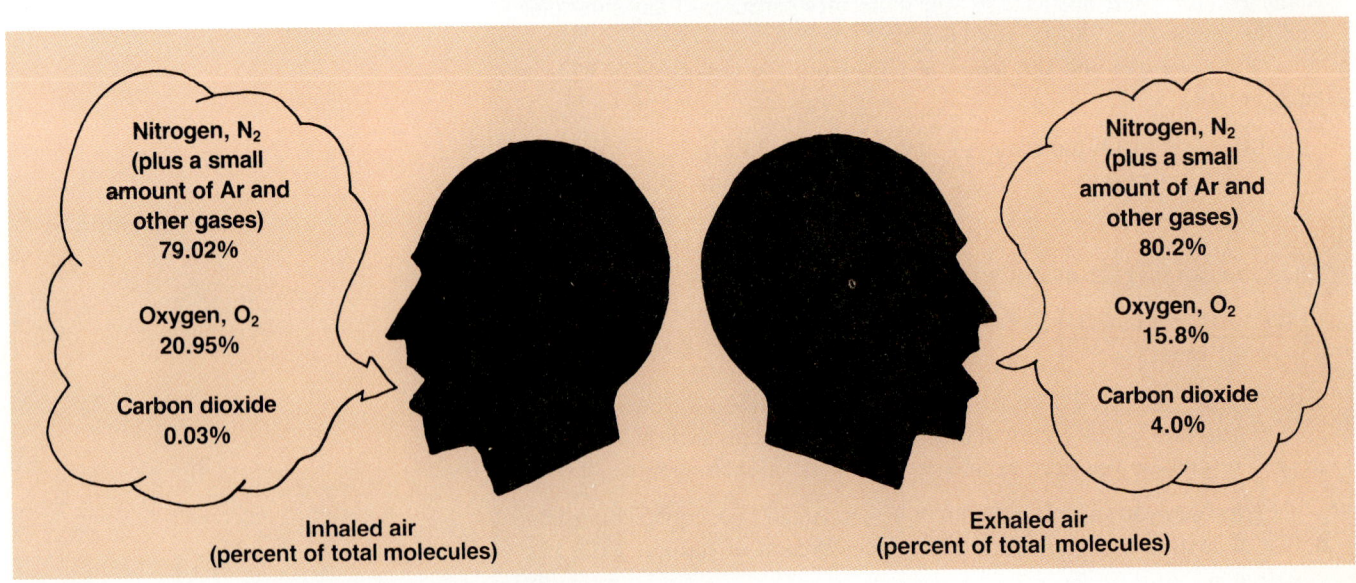

Figure VI.1 The composition of inhaled and exhaled air.

b. Assuming an average of 14 breaths per minute, how many breaths do you take per day? What factors could result in different answers? Why?
c. Say the volume of air inhaled with each breath is 500 mL. How many liters of air would you inhale each minute? Each day?
d. According to Figure VI.1, your lungs extract only a small fraction of the inhaled oxygen gas. What do you think determines the amount of oxygen used?
e. Assume that there are 250×10^{20} molecules in a 1-L sample of air. Using your answer to Question 1c, calculate how many oxygen molecules are present in the air you inhale each day? In the air you exhale each day? How many oxygen molecules do you use per day?
f. Use the result of your final calculation in Question 1e plus the number of molecules per mole (6.02×10^{23}) and the molar mass of oxygen gas to calculate the mass of oxygen you use per day.

2. Write the chemical equation for the "burning" (combustion) of 1 mol of glucose ($C_6H_{12}O_6$) to give carbon dioxide (CO_2) and water (H_2O). This will represent what occurs in the body.
   a. Given an equal number of moles of oxygen gas and glucose, which would be the limiting reactant in this combustion reaction? Why? (Refer to Unit Four, Part C.2 if you wish to review limiting reactants.)
   b. Given equal masses of oxygen gas and glucose, which would be the limiting reactant? Why?
   c. Using the equation you wrote, the mass of oxygen gas used each day found in the answer to Question 1f, and the molar masses of oxygen and glucose, calculate the mass of glucose burned each day. If glucose produces approximately 17 kJ for every gram burned, how much heat energy would be generated each day?
   d. What substance produced in your body's "burning" of glucose does not appear in Figure VI.1? How could you confirm that this substance is present in exhaled breath?

Plant life needs a continual supply of carbon dioxide—a waste product of animal respiration. Through photosynthesis in green plants (discussed in the food unit), carbon dioxide combines with water, ultimately forming more glucose and oxygen gas. Photosynthesis and respiration thus balance each other—and the concentration of oxygen gas in the atmosphere remains constant (see Figure VI.2).

The atmosphere can restore and cleanse itself if natural systems are not burdened by excessive pollution. Environmental problems arise when human activities overwhelm the natural recycling and cleansing systems. Air pollution will be explored in Parts C and D of this unit, but as a prelude, you will begin to consider this important problem in the next activity.

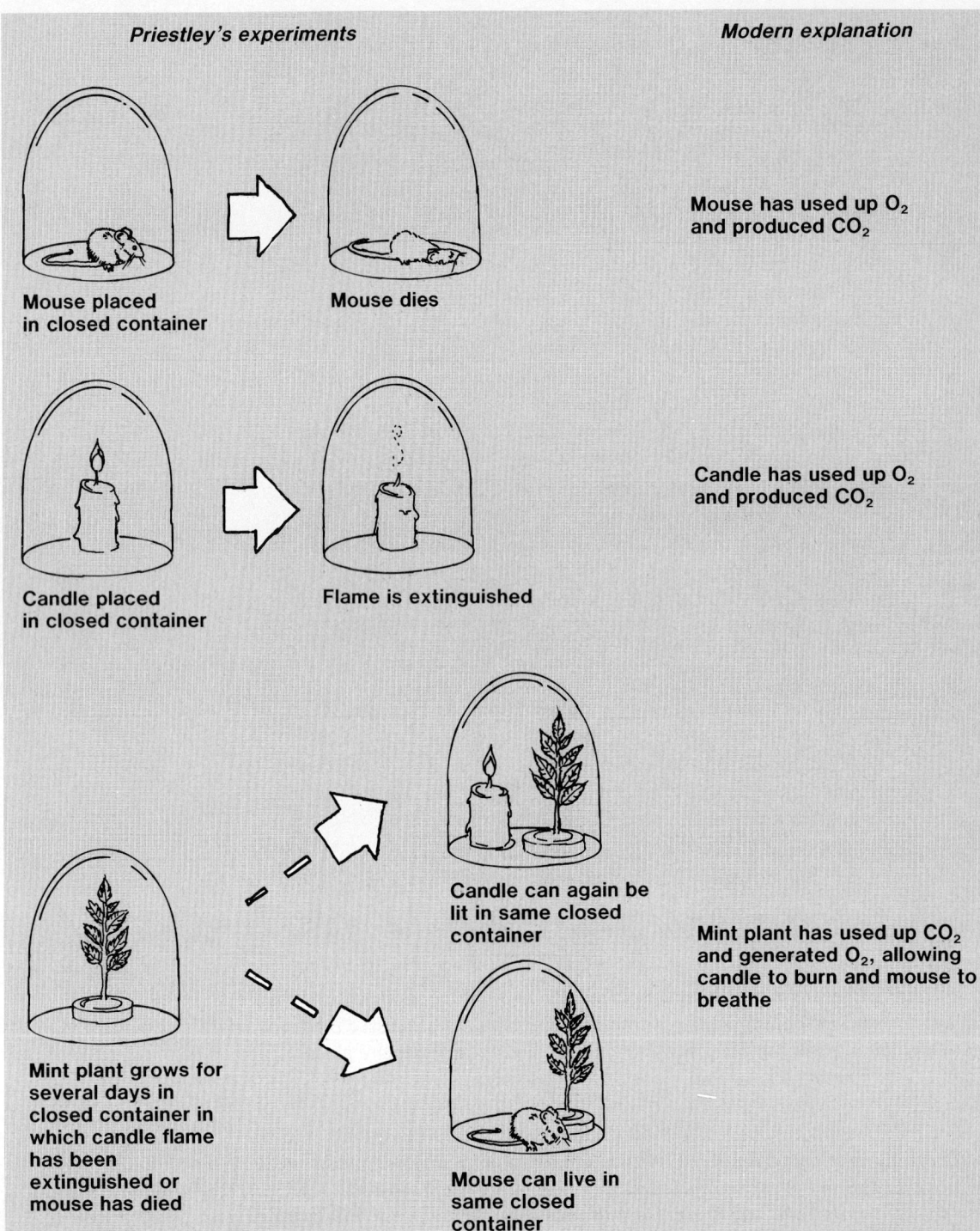

Figure VI.2 Joseph Priestley's experiments in photosynthesis and respiration. In the late 1700s, Priestley performed experiments in which he observed the effects on air of plants, animals, and combustion in various combinations. Although it was not understood until many years later, Priestley had demonstrated that plants use carbon dioxide and give off oxygen, but animals use oxygen and give off carbon dioxide.

A.4   *You Decide:* Air—Just Another Resource?

Consider these two points of view:

View A: Access to air is a fundamental human right; therefore to tax or fine individuals or industries that extract materials from or add pollutants to air is unjust.

View B: Access to air is a fundamental human right; therefore air cleanliness should be ensured by government regulations, which provide taxes or fines upon individuals or industries that lessen air quality.

Identify at least five scientific, practical, or social questions you would like answered before choosing between these positions. As you proceed through this unit, try to seek answers to the questions you have listed.

Given the potentially damaging effects of human activities on air quality, we must learn how to use the atmosphere wisely. As a first step, we must understand its chemical composition and some properties of gases. This is the focus of the next part of our study.

PART A
SUMMARY QUESTIONS

1. Describe at least two ways air is similar to and two ways it is different from other resources (such as water, minerals, and petroleum) that you have already studied.
2. Defend or refute each of these statements about gas behavior, based on the gas properties you saw demonstrated earlier in this unit.
    a. An empty bottle is not really empty.
    b. Atmospheric pressure acts only in a downward direction.
    c. Gases naturally "mix" or diffuse by moving from regions of low concentration to regions of high concentration.
    d. All colorless gases have the same physical and chemical properties.
3. As we have noted, you inhale about 14 breaths of air per minute, or about 14 kg of air per day. Based on these data, what is the mass (in grams) of a typical breath of air? If we are concerned with the chemistry involved in breathing, suggest another way the "size" of a breath of air might be expressed. Why would this be useful?

B INVESTIGATING THE ATMOSPHERE

Concern with air quality began with the discovery of fire. Fires for heating, cooking, and metalworking created air pollution problems in ancient cities. In 61 A.D., the philosopher Seneca described "the heavy air" of Rome and "the stink of the smoky chimneys thereof." The Industrial Revolution, beginning in the early 1700s, triggered increased air pollution in and around cities. These developments were accompanied by increased interest in the chemistry and physics of the atmosphere.

In the following sections you'll extend your understanding of the atmosphere, based in part on the results of the demonstrations you observed in Part A.2. This knowledge will be useful in predicting the general behavior of gases. It will help you understand how the sun's radiant energy interacts with atmospheric gases to create weather and climate. Also, this discussion sets the stage for a later discussion of air pollution. We will begin by preparing and investigating some atmospheric gases.

B.1 Laboratory Activity: Chemical Composition of the Atmosphere

Getting Ready

Air, as was shown in Figure VI.1, is composed primarily of nitrogen (N_2) and oxygen (O_2) gases together with much smaller amounts of carbon dioxide and several other gases. Each component of air has distinct physical and chemical properties. In this activity your class will generate two gases present in the atmosphere—oxygen and carbon dioxide—and examine some of their chemical properties.

Oxygen will be generated by the following reaction

$$2\ H_2O_2(aq) \xrightarrow{MnO_2} 2\ H_2O(l) + O_2(g)$$
$$\text{Hydrogen peroxide} \qquad \text{Water} \qquad \text{Oxygen gas}$$

Carbon dioxide will be generated by adding antacid tablets to water. These tablets contain a mixture of sodium bicarbonate and potassium bicarbonate ($NaHCO_3$ and $KHCO_3$) and citric acid (an organic acid present in lemons and oranges). When dissolved in water, the bicarbonate ions (HCO_3^-) react with hydrogen ions (H^+) from the citric acid to generate carbon dioxide (the "fizz"):

$$H^+(aq) + HCO_3^-(aq) \rightarrow H_2O(l) + CO_2(g)$$

Jon Jacobson

Oxygen gas extracted from air is a valuable resource.

The behavior of oxygen and carbon dioxide in the presence of burning magnesium, hot steel wool, and burning wood will be investigated. Do oxygen and carbon dioxide support the combustion of these materials?

Next, you will observe whether these gases react with limewater (a solution of calcium hydroxide, $Ca(OH)_2$). Finally, the acid-base properties of oxygen and carbon dioxide will be investigated. Acidic substances produce H^+ ions in aqueous solution, basic substances produce OH^- in aqueous solution, and neutral substances produce neither. In Unit One, Part C.6, you learned that the pH scale can be used to express the acidity or basicity of a solution. Universal indicator contains various compounds; each changes color at a different pH. (In Parts D.13 and D.14 of this unit, acids and bases are discussed further.)

If a tank of compressed nitrogen gas is available, your class also will investigate the properties listed above for nitrogen.

Read the procedures below and set up a data table in your notebook for the results of the tests on nitrogen, oxygen, and carbon dioxide. If nitrogen is not available to be tested, your teacher will give you the information needed to fill in the blanks.

Procedure

Part 1: Gas Preparation
All Groups

Label two gas-collecting bottles "Sample 1" and "Sample 2." Label three test tubes "Sample 3," "Sample 4," and "Sample 5."

Figure VI.3 Gas-generating setup.

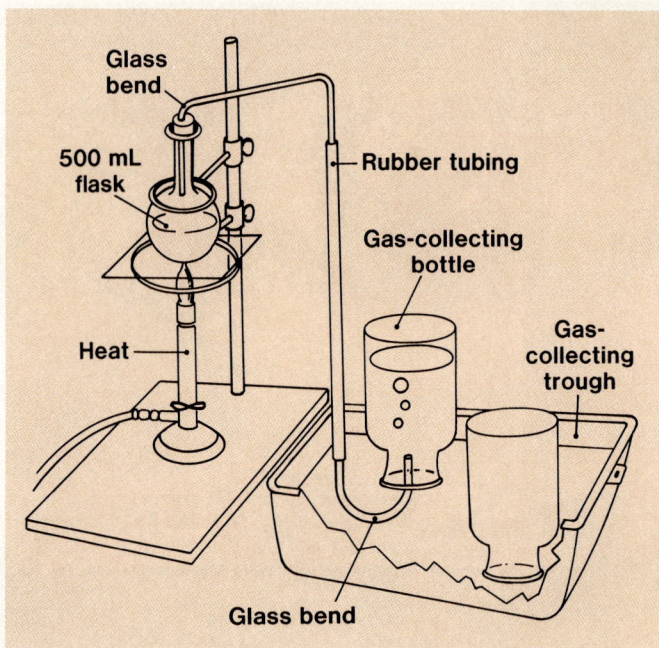

Oxygen Group

1. Set up the apparatus illustrated in Figure VI.3 for generating oxygen gas.
2. Fill the gas-collecting trough with enough tap water to cover the small shelf to a depth of about 3 cm. Fill the two gas-collecting bottles with water. Position them in the trough as your teacher directs.
3. Add 200 mL of 3% hydrogen peroxide (H_2O_2) to a 250-mL graduated cylinder. Weigh 1 g of manganese dioxide (MnO_2) on a clean paper square. (*Caution:* Avoid inhaling any MnO_2 dust.)
4. Pour the hydrogen peroxide into a 500-mL flask and carefully add the manganese dioxide to the flask.
5. Quickly place the stopper with attached glass and rubber tubing in the flask. Place the other end of the tubing under the surface of the water in the trough, but not yet under the mouth of the gas collection bottle. If the reaction is not occurring fast enough to generate a steady stream of gas bubbles, heat the mixture *gently* (but do not allow it to boil) until it reaches a constant rate of gas evolution; heat gently as you collect the gas in Step 6.
6. Once the reaction is smoothly under way, allow gas to bubble into the water for one minute. Then position the tube under the collecting bottle mouth.
7. Collect two bottles and three test tubes of the prepared gas as follows: When the first jar is filled you'll observe bubbles rising to the surface of the trough water. Hold a glass plate tightly to the mouth of the inverted jar. Remove the sealed jar from the trough. Set it upright on the table, with the glass plate covering the top. Move the tubing and fill the second bottle.

    To collect a test tube of gas, fill the test tube completely with water. Hold a finger tightly over the opening and invert the tube in the water tray. Direct the stream of gas into the inverted test tube mouth. When gas fills the tube, cover the tube mouth with your finger and remove it from the tray. Immediately close the tube with the stopper.

    If the reaction stops before you have collected the necessary volume of oxygen gas, remove the stopper from the generating flask, remove the burner and allow the flask to cool for five minutes. Then add 20 mL of H_2O_2 solution to the reaction flask.
8. After sufficient gas has been collected, remove the stopper from the generating flask before removing the heat.

Carbon Dioxide Group
1. Set up the apparatus illustrated in Figure VI.3 for generating carbon dioxide gas. The burner is not needed here.
2. Add tap water to the gas-collecting trough to cover the small shelf to a 3-cm depth. Fill two gas-collecting bottles with water. Position them in the trough as your teacher directs.
3. Place 250 mL of water in the 500 mL flask in the generating apparatus. Drop two antacid tablets into the water.
4. Quickly place the stopper with attached glass and rubber tubing in the flask. Place the other end of the tubing under the water surface in the trough, but not under either gas collection bottle.
5. Allow the gas to bubble into the trough for one minute. Then position the tube under the mouth of a collecting bottle.
6. Collect two bottles and three test tubes of the gas as described in Step 7 of the oxygen procedure. If the gas generation slows down before you are finished, add another antacid tablet to the generating flask.

Nitrogen Group
Fill two gas-collecting bottles and three test tubes with nitrogen gas from the nitrogen tank. Follow the procedure described in Steps 1, 2, 5, and 6 for the carbon dioxide group, with any modifications your teacher specifies.

Part 2: Gas Tests
Each group will perform the following on its assigned gas, and share results with the class. Record observations for each gas in your data table.

1. Light a bunsen burner. Using tongs, hold a 10–15 cm-long piece of magnesium ribbon in the flame until it ignites. (*Caution:* Do not look directly at the intense light of the burning magnesium.) Quickly remove the cover plate from Sample 1 and plunge the burning magnesium into the gas in the jar. Hold it there until you and the other class members have observed the result. Record your observations.
2. Pour about 10 mL of water into the bottom of Sample 2, quickly, with the cover pushed aside just a little. Using tongs, hold a piece of steel wool (spread out; not in a tight ball) in the burner flame until it glows red hot. Quickly plunge the steel wool into Sample 2. Hold it there, noting the result.
3. Using tongs, hold a wood splint in the burner flame until it lights. Blow out the flame, then blow on the embers so they glow red. Remove the stopper from Sample 3, and quickly plunge the glowing splint into the collected gas. Turn off the burner.
4. Add about 2 mL of limewater to Sample 4. Close the tube with the stopper and shake carefully. Note any changes in the appearance of the liquid.
5. Add two drops of universal indicator solution to Sample 5. Put the stopper back in the tube and shake carefully. Compare the results with the color chart provided with the indicator. Report the estimated pH to the class.

Questions
1. Why was it important to allow some gas to bubble through the water before you collected a sample?
2. Which gas appears to be the most reactive? The least reactive?
3. Which gas appears to support rusting and burning? This gas makes up (about) what percent of air molecules? (See Figure VI.1.) If our atmosphere contained a higher concentration of this gas, what might be some consequences?
4. Carbon dioxide is used in some types of fire extinguishers. Explain why it works.

5. Figure VI.1 suggests that human lungs expel carbon dioxide. Describe two ways you would verify that your exhaled breath contains CO_2.

What gases in addition to nitrogen, oxygen, and carbon dioxide are present in the atmosphere? Using a variety of sampling and measuring techniques, scientists have pieced together a detailed picture of the atmosphere's chemical composition. This is the subject of the next section.

B.2 A Closer Look at the Atmosphere

Most of the atmosphere's mass, and all of the weather, are located in the 10 to 12 km immediately above the Earth's surface. This region, called the **troposphere** (after the Greek word meaning "to turn over"), is the one in which we live. We will examine it first.

Continuous mixing of gases occurs within the troposphere, making its chemical composition reasonably uniform (Table VI.1). Analysis of glacial ice core samples suggests that the troposphere's composition has remained relatively constant throughout human history.

Table VI.1 *Composition of pollution-free, dry tropospheric air*

Substance	Formula	Percent of gas molecules
Major components		
Nitrogen	N_2	78.08
Oxygen	O_2	20.95
Minor components		
Argon	Ar	0.934
Carbon dioxide	CO_2	0.0314
Trace components		
Neon	Ne	0.00182
Ammonia	NH_3	0.00100
Helium	He	0.000524
Methane	CH_4	0.000200
Krypton	Kr	0.000114

NASA

The only component of the atmosphere visible from space is condensed water vapor.

In addition to the gases listed in Table VI.1, actual air samples may contain up to 5% water vapor. In most locations the normal range for water vapor is 1–3%. Other gases are naturally present in concentrations below 0.0001% (1 ppm). These include hydrogen (H_2), xenon (Xe), ozone (O_3), nitrogen oxides (NO and NO_2), carbon monoxide (CO), and sulfur dioxide (SO_2). Human activity can alter the concentrations of carbon dioxide and some trace components. It can also add new substances that may lower air quality, as we shall see.

If air is cooled under pressure, it condenses to a liquid that boils at about −185 °C. Distilling this liquid produces pure oxygen, nitrogen, and argon.

Liquid air distillation is basically the same as petroleum distillation.

Imagine that we have one mole each of oxygen, nitrogen, and argon gas. Assume we place these identical amounts in three very flexible balloons. Here is what we would observe:

- The three balloons would occupy the same volume, assuming external air pressure and temperature are the same for each.
- The three balloons would shrink to half this volume if the outside air pressure were doubled and the temperature held constant.
- The three balloons would expand to the same extent if the temperature were increased by 10 °C and the external air pressure held constant.
- Each balloon would shrink to half its original volume if half the gas were removed and neither the external air pressure nor the temperature changed.

Study has shown that a balloon filled with one mole of any other gas expands and contracts in exactly the same way, provided the temperature does not drop too low or the external pressure rise too high. All gases share these same expansion and contraction properties.

The four observations described above suggest some important conclusions about gases. One of these is this simple relationship: **Equal volumes of gases at the same temperature and pressure contain the same number of molecules.** This idea was first proposed in 1811 by an Italian chemistry teacher, Amedeo Avogadro (it is called **Avogadro's law**). If you look carefully, you can see the law at work in each observation above.

In the first three cases, each balloon contains 6.02×10^{23} molecules: one mole. In the first observation, all three balloons have the same volume at the same temperature and pressure—just as Avogadro suggested. In the second observation, the external pressure is doubled and the gas volume is half of what it was. Despite this change in conditions, all three balloons are the same size, and all contain the same number of molecules.

Now it's your turn.

YOUR TURN

VI.2 Avogadro's Law

Explain how Avogadro's law is illustrated in the third and fourth observations.

An important consequence of Avogadro's law is that all gases have the same **molar volume** if they are at the same temperature and pressure. (The molar volume is the volume occupied by 1 mol of a gas.) Scientists refer to 0 °C and 1 atm (atmosphere) pressure as **standard temperature and pressure,** or **STP**. At STP the molar volume of any gas is 22.4 L.

At STP, gas molar volume = 22.4 L/mol

The fact that all gases have the same molar volume under the same conditions simplifies our thinking about reactions involving only gases. For example, consider these equations:

$$N_2(g) + O_2(g) \rightarrow 2\,NO(g)$$
$$2\,H_2(g) + O_2(g) \rightarrow 2\,H_2O(g)$$
$$3\,H_2(g) + N_2(g) \rightarrow 2\,NH_3(g)$$

You've learned that the coefficients in such equations represent the relative numbers of molecules or moles of reactants and products. Using Avogadro's law (equal numbers of moles occupy equal volumes), you can interpret the coefficients in terms of gas volumes:

$$1 \text{ volume } N_2(g) + 1 \text{ volume } O_2(g) \rightarrow 2 \text{ volumes } NO(g)$$
$$2 \text{ volumes } H_2(g) + 1 \text{ volume } O_2(g) \rightarrow 2 \text{ volumes } H_2O(g)$$
$$3 \text{ volumes } H_2(g) + 1 \text{ volume } N_2(g) \rightarrow 2 \text{ volumes } NH_3(g)$$

The actual volumes could have any specific values, such as 1 L, 10 L, or 100 L. In the third equation, for example, we could use 300 L of H_2 and 100 L of N_2, and expect to make 200 L of NH_3 if the conversion is complete.

Rather than determining the masses of reacting gases, chemists can monitor some reactions by measuring the volumes of the gases involved. Unfortunately, similar "short cuts" do not work for reactions involving liquids and solids. There is no simple relation between moles of solids and liquids and their relative volumes.

Complete the following activity to check your understanding of the concepts discussed in this section.

YOUR TURN

VI.3 Molar Volume and Reactions of Gases

1. What volume would be occupied at STP by 3 mol of $CO_2(g)$?
2. In a certain reaction, 2 mol of $NO(g)$ reacts with 1 mol of $O_2(g)$. How many liters of $O_2(g)$ would react with 4 L of NO gas?
3. Poisonous carbon monoxide gas forms when fossil fuels such as petroleum burn in the presence of insufficient oxygen gas. The CO is eventually converted into CO_2 in the atmosphere. Automobile catalytic converters are designed to speed up this conversion:

    Carbon monoxide gas + Oxygen gas → Carbon dioxide gas

    Write the equation for this conversion. How many moles of oxygen gas would be needed to convert 50 mol of carbon monoxide to carbon dioxide? How many liters of oxygen gas would be needed to react with 1120 L of carbon monoxide? (Assume both gases are at the same temperature and pressure.)

4. In the third unit on petroleum, you studied the combustion of octane in air:

    $$2\,C_8H_{18}(l) + 25\,O_2(g) \rightarrow 16\,CO_2(g) + 18\,H_2O(g) + \text{Energy}$$

    You now know that air is a mixture of about 21% O_2 and 79% N_2 molecules (with small amounts of argon and other gases). How many moles of nitrogen gas pass through the engine with each 25-mol sample of oxygen gas? How many grams of nitrogen? At STP, how many liters of nitrogen gas would this represent?

Now that you have explored the composition and properties of gases at the bottom of the sea of air in which we live, let's continue our study of the atmosphere by swimming up to higher levels.

B.3 *You Decide:* How Do Atmospheric Properties Vary with Altitude?

If you were to dive from the ocean's surface downward you would encounter increasing pressures and decreasing temperatures. Creatures deep in the ocean live in a much different environment than those living near the surface. Similarly, the environment at sea level is quite different from that at higher altitudes. From early explorations of the atmosphere in hot-air and lighter-than-air balloons to current journeys through the atmosphere and beyond, scientists have obtained information regarding conditions in the atmosphere.

Picture yourself in a craft designed to fly from the Earth's surface to the farthest regions of our atmosphere. The instruments in your imaginary craft have been set to record altitude, air temperature, and pressure readings. One-liter gas samples can be taken at specific heights. You'll receive a report on the mass, number of molecules, and composition of each sample.

As the craft rises, the composition of the air remains essentially constant at 78% N_2, 21% O_2, 1% Ar, plus trace elements. At 12 km you notice that you are above the clouds. The tallest mountains are far below. The sky is a light blue color, and the sun shines brilliantly. The craft is now above the region where commercial aircraft fly and weather develops.

Above 12 km, air composition is about the same as at lower altitudes, except there is more ozone (O_3). You also notice that the air is quite calm, unlike the turbulent air you encountered below 12 km.

Air samples taken at 50–85 km contain relatively few particles. Those present are charged species such as O_2^+ and NO^+. Above 200 km, your radar detects various communication satellites in orbit.

Your aircraft now returns to Earth, and you turn to analysis of the data you collected.

Jon Jacobson

What limits the height to which this type of balloon can rise?

Plotting the Data

Table VI.2 presents the data recorded during the flight. The last two columns provide comparisons of equal-volume (1-L) samples of air at different altitudes.

Table VI.2 *Atmospheric Data*

Altitude (km)	Temperature (°C)	Pressure (mmHg)[a]	Mass of 1-L sample (g)	Total molecules (in a 1-L sample)
0	20	760	1.225	250 × 10^{20}
5	−12.5	407	0.736	153 × 10^{20}
10	−45	218	0.414	86 × 10^{20}
12	−60	170	0.377	77 × 10^{20}
20	−53	62.4	0.089	19 × 10^{20}
30	−38	17.9	0.018	4 × 10^{20}
40	−18	5.1	0.004	0.80 × 10^{20}
50	2	1.5	0.001	0.22 × 10^{20}
60	−26	0.42	0.0003	0.06 × 10^{20}
80	−87	0.03	b	b

[a]Millimeters of mercury (mmHg), a common pressure unit, is discussed in Part B.4.
[b]Values too small to be measured.

Prepare two graphs. One should be a plot of temperature vs. altitude; the other, pressure versus altitude. Arrange your axes to use the full sheet of paper for each graph. The x axis scale (altitude) for each graph should range from 0 to 100 km. The y axis scale (temperature) on the first graph should range from -100 °C to $+40$ °C. The y axis scale (pressure) in the second graph should run from 0 to 780 mmHg, a common unit for pressure (see Part B.4).

Plot the data collected and connect the points with the smoothest line possible. (Note that a "line" may be straight or curved.) Use these graphs and the knowledge gained from your flight to answer the following questions.

Questions

1. Compare the ways air temperature and pressure change with increasing altitude. Which follows a more consistent pattern? Try to explain this behavior.
2. Would you expect barometric pressure to rise or fall if you traveled from sea level to
   a. Pike's Peak (4270 m above sea level)?
   b. Death Valley (85 m below sea level)? Explain your answers.
3. When two rubber plungers (like those plumbers use) are pressed together, it is difficult to separate them (you may wish to try this). Why? Would it be easier or harder to separate them at the top of a mountain? Explain your answer.
4. How do your values of mass and molecules per liter of air change with increasing altitude? If you plotted these values, what would be the appearance of the lines on the graph?
5. Scientists often consider the atmosphere in terms of layers: **troposphere** (nearest earth), **stratosphere, mesosphere,** and **thermosphere** (outermost layer). What flight data evidence supports the idea of such a "layered" atmosphere? Mark the graphs with horizontal lines to indicate where you think various layer boundaries are.

In previous sections we have used the term "air pressure." In the next section we'll clarify what a scientist means by pressure.

B.4 Air Pressure

In everyday language, we speak of being pressured, meaning we feel forced to behave in certain ways. The greater the pressure, the less "space" we feel we have. To scientists, pressure also refers to force and space, but in physical ways.

Pressure represents the force applied to one unit of surface area:

$$\text{Pressure} = \frac{\text{Force}}{\text{Area}}$$

To get a better idea of this, consider a lead block with a 7.0-in. length, 3.0-in. width, and 2.0-in. height. Such a block would have a weight (exert a force downward) of 17 lb. Although the block's weight remains constant, the pressure it exerts depends upon which side the block is resting on (see Figure VI.4). The calculations of the pressure exerted in each position are

Position 1: $\dfrac{17 \text{ lb}}{7.0 \text{ in.} \times 3.0 \text{ in.}} = 0.81 \text{ lb/in.}^2$

Position 2: $\dfrac{17 \text{ lb}}{7.0 \text{ in.} \times 2.0 \text{ in.}} = 1.2 \text{ lb/in.}^2$

Position 3: $\dfrac{17 \text{ lb}}{2.0 \text{ in.} \times 3.0 \text{ in.}} = 2.8 \text{ lb/in.}^2$

Let's consider some practical applications of a scientific view of pressure.

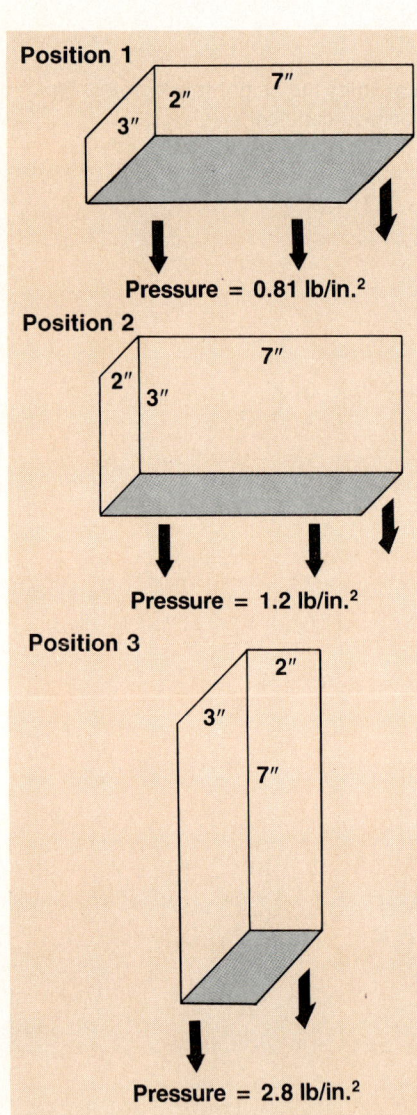

Figure VI.4 The pressure exerted by the lead block depends upon which side is down. For the same block, the pressure is greatest for the side with the smallest area, in this case, the 3 in. × 2 in. side.

CHEM QUANDARY

1. During the 1950s, stiletto-heeled shoes were quite popular. Floors across the nation became scarred with tiny dents. When tennis shoes or shoes with broader heels were worn, no such problem developed. Why? (See Figure VI.5.)
2. To avoid excessive record wear, the tone arm on a turntable should be adjusted so that the stylus (needle) has a very light tracking weight. Why does this help extend the life of records (and of the stylus)?

Figure VI.5 The pressure a shoe heel exerts depends upon its area.

1 atm = 14.7 lb/in.² = 760 mmHg

How do lead blocks, high-heeled shoes, and record styluses relate to the atmosphere? As this unit's opening demonstrations showed, the atmosphere exerts a force on every object within it. On a typical day, at sea level, air exerts a pressure of about 14.7 lb on every square inch of your body. This pressure is equal to one atmosphere. But there is yet another way in which pressure is expressed. The pressure data on the imaginary flight were given in **mmHg (millimeters of mercury)**. A weather report may include "the barometric pressure is 30 inches of mercury." Air pressure can also be expressed as the height of a mercury column. Why? A simple demonstration will clarify matters.

Fill a graduated cylinder (or soft drink bottle) completely with water. Cover it with your hand and invert it into a container of water. What happens to the water level inside the cylinder? What force supports the weight of the column of water in the cylinder?

Now imagine repeating this with a taller cylinder, and again with an even taller cylinder. If the cylinder were made taller and taller, at a certain height water would no longer fill the cylinder to the top when it is inverted into a container of water. Why?

This experiment was first performed in the mid-1600s. The researchers discovered that one atmosphere of pressure could support a column of water no taller than 33.9 ft (10.3 m). If one tries the experiment with even taller cylinders, the water drops to the 10.3-m level, leaving a partial vacuum above the liquid. Obviously, such a tall device would be rather awkward to handle. Scientists decided to replace the water with mercury, which is 13.6 times more dense. The resulting mercury barometer (see Figure VI.6) is shorter than the water barometer by a factor of 13.6.

In the following sections we will study the effects of pressure and temperature changes on gas volume. This knowledge will help us understand the gases in our bodies, those trapped in rocks on Earth, and those in the atmosphere through which jetliners and spacecraft travel. We'll consider the effect of changes in pressure first.

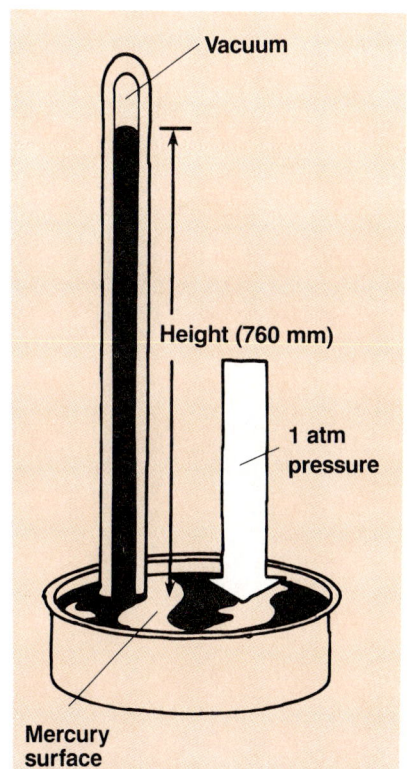

Figure VI.6 A mercury barometer. The atmosphere at sea level will support a column of mercury 760 mm high. The pressure unit "atmospheres" is thus related to pressure in millimeters of mercury, 1 atm = 760 mmHg.

B.5 Boyle's Law: How to Put the Squeeze on a Gas

Unlike the volume of a solid or a liquid, the volume of a gas can easily be changed. Applying pressure to a sample of gas decreases its volume—it is "compressed." By compressing a gas, a greater mass of the gas can be stored in a given container. Tanks of compressed gas are quite common. For example, propane tanks are used by campers, mobile home owners, and people whose homes are in rural areas; welders buy tanks of oxygen and acetylene gas; and hospitals use tanks of pure oxygen gas for patients with breathing problems. Tanks of gas are often used in chemistry laboratories.

Try the following *Chem Quandary*.

CHEMQUANDARY

A chemistry teacher notices a large difference in the price of hydrogen gas sold by two companies. Company A offers a 1-L cylinder of hydrogen gas for $8; Company B offers a 1-L cylinder of the gas for $15. The teacher discovers that Company B offers a better bargain. Explain how this could be.

Atmospheric air undergoes compression and expansion, changing with altitude (as you observed in your imaginary journey) and with weather conditions. Pressure tends to fall before a storm, and to rise as the weather clears.

How much does gas volume decrease with a given increase in pressure? You already know the answer to this. In Part B.2 it was the second of the four observations on the behavior of gases in balloons. When the pressure on the balloons was doubled, their volume decreased to half the original value.

This change in volume illustrates a relationship common to all gases. It is called **Boyle's law,** after the 17th-century English scientist who first proposed it. One way to state Boyle's law is the following: at a constant temperature, the product of the pressure and the volume of a given gas sample is a constant. Suppose we have 12 L of gas under 2 atm pressure (2 atm × 12 L = 24 L atm). Doubling the pressure to 4 atm will halve the volume to 6 L (4 atm × 6 L = 24 L atm). Note that 2 atm × 12 L = 4 atm × 6 L. Such relationships hold for any change in either the volume or pressure. If the volume is decreased to 4 L, the pressure will increase to 6 atm (2 atm × 12 L = 6 atm × 4 L).

Boyle's Law: at constant T, P·V = constant for a fixed amount of gas.

The general mathematical statement of Boyle's law is

$$P_1V_1 = P_2V_2$$

where P_1 and V_1 are the original pressure and volume of a gas, and P_2 and V_2 are the new pressure and volume at the same temperature. Plotting a series of pressure and volume values for any gas sample gives a curve similar to that in Figure VI.7.

Boyle's law allows predictions to be made of changes in the pressure or volume of a gas sample at constant temperature when any three of the values P_1, V_1, P_2, or V_2 are known. Consider the following example:

A tank of gas used in a chemistry laboratory may have a volume of 1.0 L and contain gas at a pressure of 56 atm. What volume would the gas from such a tank occupy at 1.0 atm (and the same temperature)?

We use Boyle's law to solve the problem in the following way:

1. Identify the starting and final conditions:

   $P_1 = 56$ atm $P_2 = 1.0$ atm
   $V_1 = 1.0$ L $V_2 = ?$

2. Rearrange the equation $P_1V_1 = P_2V_2$ to solve for the unknown, in this case, V_2 (the volume after the pressure decrease):

$$V_2 = \frac{P_1V_1}{P_2}$$

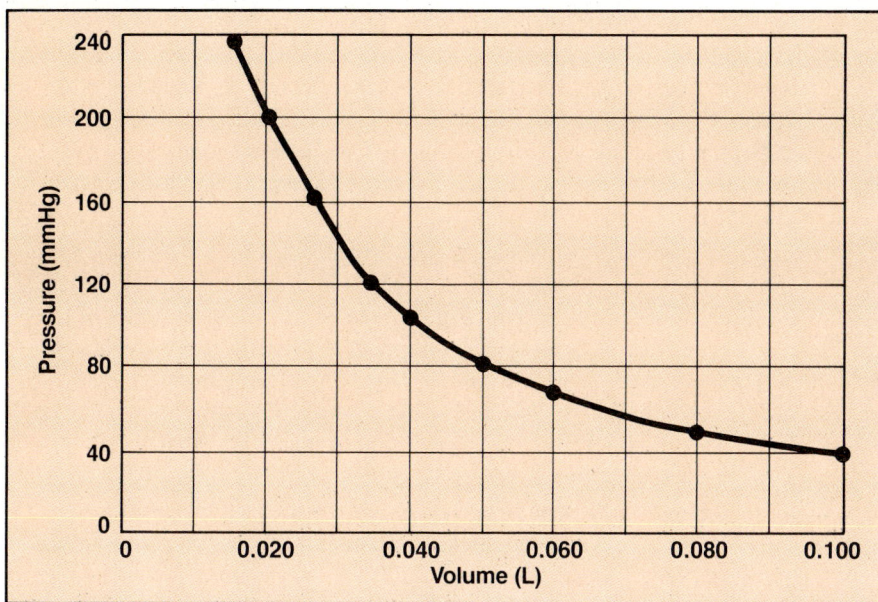

Figure VI.7 Boyle's law: The volume of a gas sample, maintained at constant temperature, is inversely proportional to its pressure. Hence the product of P × V is constant. A plot of pressure vs. volume for any gas sample at constant temperature will be similar to this one.

3. Substitute the values into the equation and solve for the unknown:

$$V_2 = \frac{(56 \text{ atm})(1.0 \text{ L})}{1.0 \text{ atm}} = 56 \text{ L}$$

A reasoning method can be used to solve the same problem:

If the pressure decreases from 56 to 1.0 atm, the volume will increase by a proportional amount. Therefore, to find the new, larger volume, the known volume must be multiplied by a pressure ratio larger than one:

$$1.0 \text{ L} \times \frac{56 \text{ atm}}{1 \text{ atm}} = 56 \text{ L}$$

If the gas pressure increases, then the volume must decrease. (This assumes that the temperature remains constant.) In this case the pressure ratio used must be less than one. Similar reasoning applies to problems in which the initial and final volume are known and the final pressure is to be found.

Work the following problems, which show some common applications of Boyle's law.

YOUR TURN

VI.4 Pressure-Volume Relationships

1. Explain each of these statements:
   a. Even when provided ample oxygen gas to breathe, aircraft passengers become uncomfortable if the cabin loses its pressure. (*Hint:* What will gases in the human body do if the body is exposed to a sudden drop in air pressure?)
   b. Bottles of carbonated soft drink "pop" when opened.
   c. Tennis balls are sold in pressurized cans. (*Hint:* Each tennis ball contains gases at elevated pressure to give it good bounce.)
   d. When climbing a mountain or ascending a tall building in an elevator, your ears may "pop."

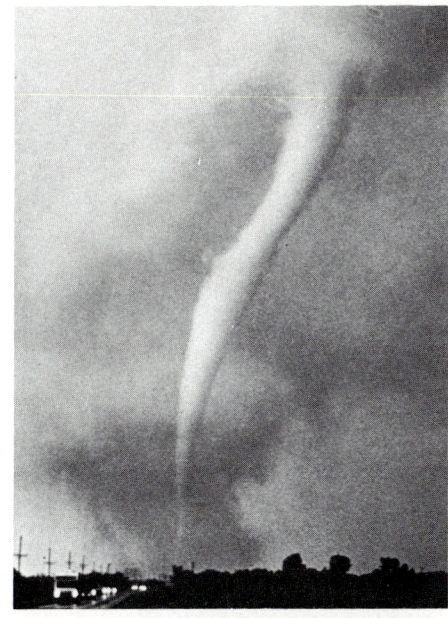

National Severe Storms Laboratory

Dramatic changes in air pressure are associated with the passage of a tornado.

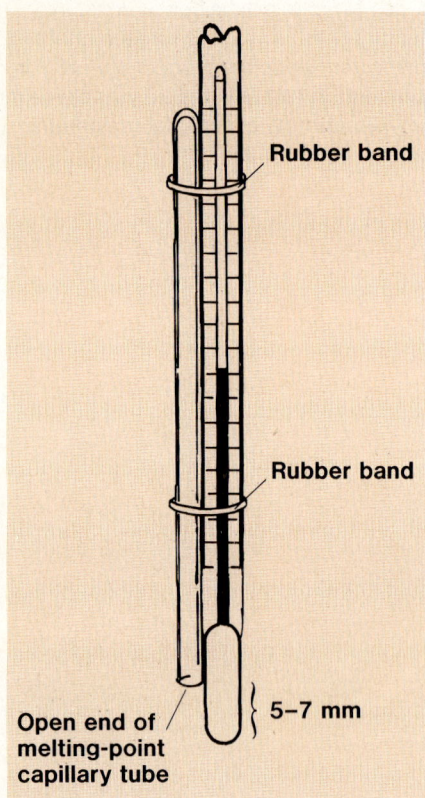

Figure VI.8 Apparatus to study variation of gas volume with temperature.

2. You can buy helium in small aerosol cans to inflate party balloons. The label on such a can states that it contains "about 0.25 cubic feet (or 7080 mL) of helium." Assume that this volume refers to the volume of helium at 1 atm pressure (760 mmHg). If the volume of the can is 492 mL, what is the gas pressure inside the can (at the same temperature)?

3. Suppose that on a hot, sticky spring afternoon, a tornado passes near your high school. The air pressure inside and outside your classroom (volume = 430 L) is 760 mmHg before the storm. At the peak of the storm, pressure outside the classroom drops to 596 mmHg. To what volume would the air in the room try to change to equalize the pressure difference between the inside and the outside? (Assume no change in air temperature.) Why is it a good idea to open the windows slightly as such a storm approaches?

4. Automobile tire pressure is usually measured with a gauge that reports the difference in pressure between the air in the tire and atmospheric pressure. Thus, if the tire gauge reads 30 lb/in.^{2} on a day when the atmospheric pressure is 14 lb/in.^{2}, the total tire pressure is 44 lb/in.^{2}. What volume of air from the atmosphere (at 14 lb/in.^{2}) would be needed to fill a 40.0-L tire to a gauge reading of 30 lb/in.^{2}? Why does pumping up an automobile tire by hand take such a long time?

In all these problems, we assumed that gas temperature remained constant. How does temperature influence gas volume if the pressure remains constant? The answer to this question is important because many reactions in laboratory work, in chemical processing, and all those in cooking are done at roughly constant atmospheric pressure. Did you ever wonder why cakes and bread rise as they bake? Read on and you'll find out.

B.6 Laboratory Activity: Temperature-Volume Relationship for Gases

Getting Ready
Most forms of matter expand when heated and contract when cooled. As you know, gases expand and contract to a much greater extent than do either solids or liquids.

In this activity you will investigate how the temperature of a gas influences its volume when pressure remains unchanged. To do this, you will raise the temperature of a thin glass tube containing a trapped air sample, and then record volume changes as the air sample cools.

Procedure
1. Tie a capillary tube to the lower end of a thermometer using two rubber bands (Figure VI.8). The open end of the tube should be placed closest to the thermometer bulb and 5–7 mm from the bulb's tip.
2. Immerse the tube and thermometer in a hot oil bath that has been prepared by your teacher. Be sure the entire capillary is immersed in the oil. Wait for your tube and thermometer to reach the temperature of the oil (approximately 130 °C). Record the temperature of the bath.

3. After your tube and thermometer have reached constant temperature, lift them so that about three-quarters of the capillary tube is elevated out of the oil bath. Pause here for about three seconds to allow some oil to rise into the tube. Then quickly carry the tube and thermometer (on a paper towel to avoid dripping) back to your desk.

4. Lay the tube and thermometer on a paper towel on the desk. Make a reference line on the paper at the sealed end of the melting-point tube. Also mark the upper end of the oil plug (Figure VI.9). Alongside this mark write the temperature at which the air column had this length.

5. As the temperature of the gas sample drops, make at least six marks representing the length of the air column trapped above the oil plug; write the corresponding temperature next to each mark. Allow enough time so that the temperature drops by 80 to 100 degrees. Since the tube has a uniform diameter, length serves as a relative measure of the gas volume.

6. When the thermometer shows a steady temperature (near room temperature), make a final length and temperature observation. Discard the tube and the rubber bands according to your teacher's instructions. Wipe the thermometer dry.

7. Measure and record (in centimeters) the marked lengths of the gas sample.

8. Prepare a sheet of graph paper for plotting your volume (length) and temperature data. The vertical scale should range from 0 to 10 cm; the horizontal scale should include values from −350 °C to 150 °C. Label your axes and arrange the scales so the graph fills nearly the entire sheet. Plot the temperature-length data. Draw the best straight line through the plotted points. Using a dashed line, extend this straight graph line so it intersects the *x* axis. Turn in one copy of your data; keep another copy for the questions that follow.

Questions

1. What was the total change in the length of your sample of gas from the first reading to the last? What was the corresponding overall change in temperature?

2. Use your Question 1 answers to find the change in length for each degree of temperature change. For instance, if the length decreased by 5.0 cm while the temperature dropped 100 °C, the ratio would be

$$\frac{5.0 \text{ cm}}{100 \text{ °C}} = 0.050 \text{ cm per °C}$$

3. Use the value calculated in Question 2 to find the gas temperature that would correspond to a sample length of 0 cm.

4. Compare the temperature found in Question 3 with the temperature noted at the point where your extended graph line intersected the horizontal axis. Both values represent estimates of the gas temperature required to shrink its volume to "zero." (This assumes, of course, that the temperature-volume trend continues to low temperatures.) Which is the better estimate of this "zero gas volume" temperature? Why?

5. If you were to continue cooling the trapped gas sample, do you think the gas would finally reach "zero volume"? Why?

Experiments similar to the one you just performed were first completed in the late 1700s. They resulted in some new ideas, including a more useful temperature scale.

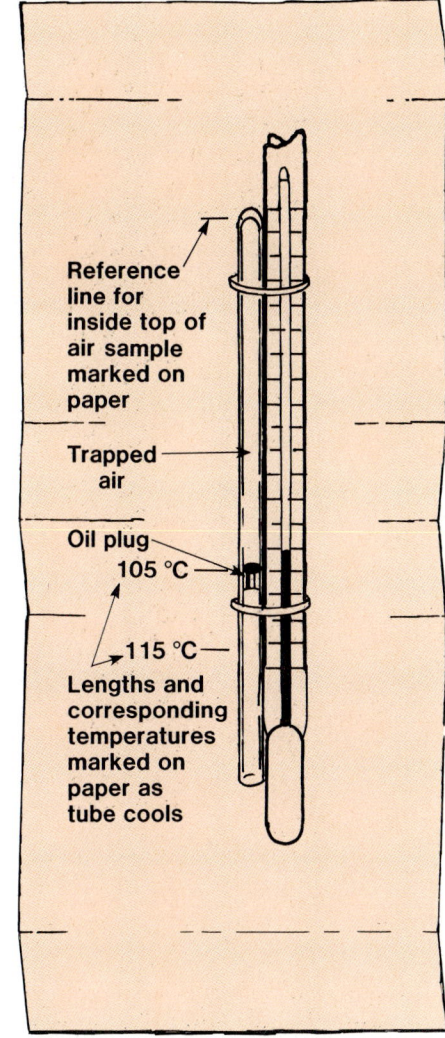

Figure VI.9 Marking the length and temperature of the air sample. The air trapped in this tube has cooled from 115 °C to 105 °C and shrunk as shown.

B.7 *You Decide:* A New Temperature Scale

Studies of changes in gas volume caused by temperature changes at constant pressure were conducted in the 1780s by French chemists (and hot-air ballooning enthusiasts) Jacques Charles and Joseph Gay-Lussac. Some data illustrating their observations are shown in Figure VI.10. The plots for different gases and different sample sizes begin at different points, but if extended to the x axis (an extrapolation), all meet at the same temperature.

Clearly, there is a simple relationship between gas volume and temperature (at constant pressure). However, expressing this mathematically proved difficult because of the need to deal with temperatures that are negative numbers on the Celsius scale. Lord Kelvin (William Thomson, an English scientist), solved the problem by proposing a new temperature scale.

Let's see if the pooled data found by your class in the just-completed lab activity will reveal what Kelvin discovered. Answer the following questions:

1. At what temperature does your extended graph line intersect the x axis?
2. What would be the volume of your gas sample at this temperature?
3. Why is this volume only theoretical?

Now renumber the temperature scale on your graph, assigning the temperature at which the graph intersects the x axis the value zero. The new scale expresses temperature in kelvin units (K), and is sometimes referred to as the kelvin temperature scale. One kelvin is the same size as one degree Celsius, but the zero point of the scale has been changed so that kelvin temperatures are all positive numbers. Zero kelvins is the lowest temperature theoretically possible: absolute zero.

4. Based on your graph, what temperature in kelvins would correspond to 0 °C, the freezing point of water? What temperature would correspond to 100 °C, the boiling point of water at one atmosphere pressure?
5. How do you convert a temperature in degrees Celsius units to its corresponding temperature in kelvins? How would you convert a temperature in kelvins to its corresponding value in degrees Celsius?

The kelvin is the SI unit for temperature.

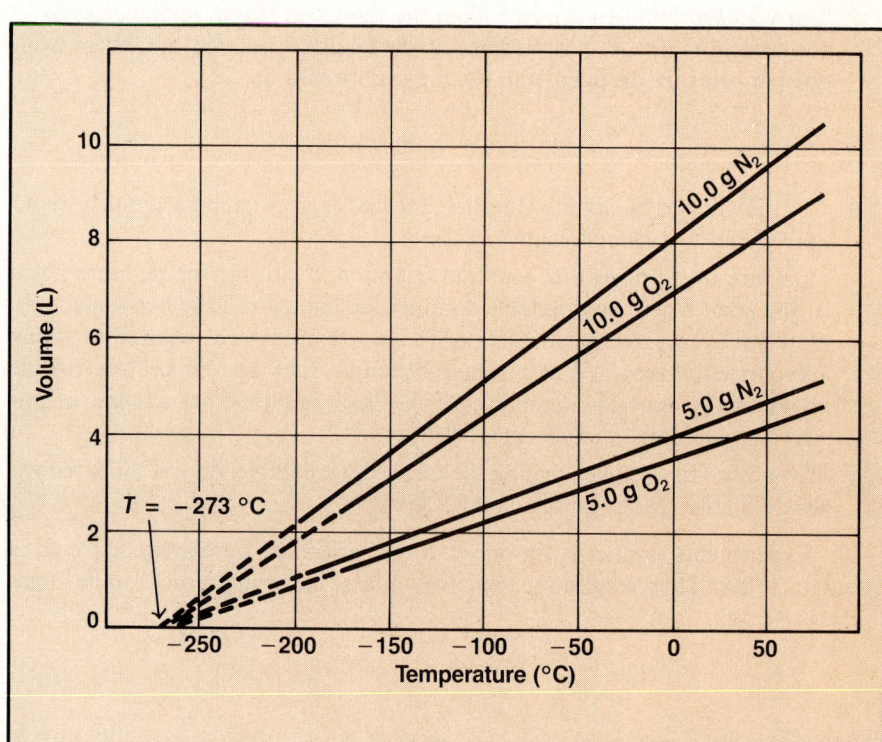

Figure VI.10 Temperature–volume measurements of various gas samples at 1 atm pressure. Extrapolation has been made for temperatures below liquefaction (O_2 is −183 °C, N_2 is −196 °C).

The kelvin temperature scale led to a simple temperature-volume relationship for gases. Doubling the kelvin temperature of a gas sample causes its volume to double, provided external pressure remains constant. Halving the kelvin temperature causes the gas volume to decrease by half, and so on. Such relationships are summarized in **Charles' law:** At constant pressure, the kelvin temperature divided by the volume of a given gas sample is a constant.

Charles' law: At constant P, $\frac{T}{V}$ = constant for a fixed amount of gas.

CHEMQUANDARY

Knowing how gases behave when temperature increases at constant pressure, answer the following questions:

a. Why does bread rise when baked? (*Hint:* $CO_2(g)$ is produced by yeast action.)
b. Is hot air more dense or less dense than cold air?
c. Why do hot air balloons rise?
d. If you can install only one thermostat in a two-story house, should it be placed on the first or second floor?

Try the following gas behavior problems using the approach you prefer. In each problem identify the variable (pressure or volume) that is assumed to be constant.

YOURTURN

VI.5 Temperature-Volume-Pressure Relationships

1. If the kelvin temperature of a fixed volume of gas increases to three times its original value, what will happen to the pressure of the gas? Give an example of such a situation.
2. If a sample of gas is cooled and contracts at a constant pressure to one-fourth of its initial volume, what change in its kelvin temperature must have occurred? Give an example of such a situation.
3. Explain why car owners in severe northern climates often add air to their tires in winter and release air from the tires when summer arrives.
4. When the volume of a gas sample is reported, the pressure and the temperature must also be given. Why? This is not necessary for liquids and solids. Why?
5. Use gas laws to explain each of the following situations:
   a. A rising weather balloon encounters decreasing pressure and decreasing temperature. It expands as it rises.
   b. Thunderheads towering high into the sky form where warm, moist air and cold air masses meet.
   c. The Earth's weather is most changeable in the middle lattitudes in the northern and southern hemispheres.

Why do all common gases at normal atmospheric conditions behave in accord with Boyle's law and Charles' law? In the next section an explanation will be developed.

B.8 In Ideal Cases, Nature Behaves Simply

By the early 1800s, scientists had discovered the gas laws you have just explored. Balloon flights also provided valuable information on the composition and structure of the atmosphere. However, explanations for why gases behave so consistently and similarly were somewhat sketchy.

Since the 17th century, scientists have pictured gases as composed of tiny particles separated by great distances. But not until the early 19th century did the atomic theory lay the groundwork for understanding gas behavior. Bit by bit, scientists pieced together the **kinetic molecular theory** of gases.

To understand the theory, you must first understand the concept of **kinetic energy.** Kinetic energy is simply energy possessed by a moving object. It is related to both the mass and the velocity of that object. At a given velocity, a more massive object has greater kinetic energy than a less massive object. And a given object has greater kinetic energy as its velocity increases.

The kinetic molecular theory of gases is based on the following postulates:

A postulate is an accepted principle used as the basis for reasoning or study.

- Gases consist of tiny particles (atoms or molecules) that have negligible size compared with the great distances separating them.
- Gas molecules are in constant, random motion. They often collide with each other and with the walls of their container. Gas pressure is the result of molecular collisions with the container walls.
- Molecular collisions are elastic. This means that although individual molecules may gain or lose kinetic energy, there is no net (or overall) gain or loss of kinetic energy from these collisions.
- At a given temperature, molecules in a gas sample have a range of kinetic energies. However, the average kinetic energy of the molecules is constant and depends only on the kelvin temperature of the sample. In samples of different gases at the same temperature, the average kinetic energy of the molecules is the same. As the temperature increases, so do the average velocities and kinetic energies of the molecules.

An analogy may be helpful in understanding the molecular picture presented by the kinetic molecular theory.

YOUR TURN

VI.6 Kinetic Molecular Theory: An Analogy

1. Imagine that people dancing in an enclosed space are a two-dimensional representation of gas molecules bouncing around a container (that is, they move across the floor, but not up to the ceiling and down). Identify which variable (volume, temperature, pressure, or number of molecules) is most like the following:
    a. The number of dancers
    b. The size of the room
    c. The tempo of the music
    d. The number and force of collisions between dancers.
2. Which law (Boyle's, Charles', Avogadro's or other) best applies to each of these situations?
    a. The tempo of the music and the number of dancers remain constant, but a partition is closed, leaving less space for dancing. (What will this do to the number and force of collisions?)

James L. Shaffer

A model for molecules in motion.

> b. The size of the room and number of dancers are kept constant, but the tempo of the music is increased. (What will this do to the number and force of collisions?)
> c. The room size and the tempo of music are kept constant, but the number of dancers is increased. (How will this affect the number and force of collisions?)

Other analogies sometimes used to help picture gases at the molecular level include a room swarming with tiny gnats, or one full of bouncing superballs. Although helpful in understanding molecular behavior, such models fail to represent these characteristics:

- The extremely small size of the molecules relative to the total volume of a sample of gas. Less than one ten-thousandth of the volume of nitrogen gas is taken up by particles. The rest of the space is empty.
- The extremely high velocities of the molecules. Under normal conditions of pressure and temperature, most molecules move at speeds greater than 1700 km/h (1100 mph).
- The extremely high frequency of collisions.

Because of these differences, the actual behavior of gases can be represented precisely only through mathematical expressions.

Complete the following exercises to check your understanding of the kinetic molecular theory of gases.

YOUR TURN

VI.7 Gas Molecules in Motion

Use the kinetic molecular theory to explain the following observations regarding gases. Where appropriate, discuss such factors as the kinetic energies, spaces between molecules, and collisions of molecules. A sketch of molecules in motion may also be helpful for some items. An example is provided for you.

Increasing the volume of a gas sample will decrease its pressure if the temperature is constant.

Increasing the volume of a sample of gas molecules gives the molecules more room to move. Thus molecules must travel farther (on average) before colliding with the container walls. Fewer molecular collisions with the walls in a given period of time means that the pressure will decrease.

1. Decreasing the volume of a gas sample held at constant temperature causes the gas pressure to increase.
2. At constant pressure, the volume of a gas changes when the temperature changes.
3. At constant volume, the pressure of a gas changes when the temperature changes.
4. The atmosphere exerts a pressure on our bodies, yet this pressure doesn't crush us.
5. a. Filled balloons eventually leak even if they are tightly sealed.
   b. Helium balloons leak faster than those inflated with human breath.
6. Gases spread out from regions of greater concentration (or higher pressure) to regions of lesser concentration (or lower pressure) until they become uniformly mixed. This process is called **diffusion.** Explain these examples of gaseous diffusion:
   a. Smelly gaseous pollutants can be detected several kilometers from their source even on days with no noticeable breeze.
   b. The percentage of different gases in the troposphere is essentially constant (except for H_2O) at all geographic locations and altitudes.

Gases that behave as the kinetic molecular theory predicts are referred to as **ideal gases.** At very high pressures or very low temperatures, gases do not behave ideally. Boyle's law and Charles' law no longer accurately describe gas behavior under such conditions. At low temperatures the molecules move more slowly and the weak attractions between them may become so important that the gas will condense into a liquid. At high enough pressures, if the temperature is not too high, the molecules are forced so close together that these forces of attraction may again cause a gas to condense. However, under conditions normally encountered in our atmosphere, most gases behave close to ideally and their behavior is well explained by the kinetic molecular theory.

Now that you understand how chemists account for some of the properties of gases, you are ready to explore further how the gases in our atmosphere interact with the sun's radiant energy to create weather and climate.

PART B
SUMMARY QUESTIONS

1. Oxygen gas is essential to life as we know it. Our atmosphere contains approximately 21% oxygen. Would a higher percentage of atmospheric oxygen be desirable? Explain.
2. Explain the following reaction using a molecular picture, a chemical equation, and words:

    1 L Hydrogen gas + 1 L Chlorine gas → 2 L Hydrogen chloride gas

3. Does Avogadro's law apply to liquids and solids? Explain your answer.
4. In terms of the pressure exerted at various depths, how is our atmosphere similar to the oceans?
5. Explain why a suction cup hook can support the weight of a small object.
6. Explain why life does not naturally exist above the troposphere.
7. Solve these problems. As part of each answer, identify which gas variable was assumed constant and which gas law applied:
    a. A small quantity of the inert gas argon (Ar) is added to light bulbs to reduce the vaporization of tungsten atoms from the filament. What volume of argon at 760 mmHg of pressure is needed to fill a 0.210-L light bulb at 1.30 mmHg pressure?
    b. A tank at 300 K contains 0.285 L of gas at 1.92 atm pressure. The maximum pressure the container is capable of withstanding is 5.76 atm. Assuming that doubling the kelvin temperature causes the pressure to double, at what temperature will the tank burst? Would it explode in a 1275 K fire?
8. Convection, the process whereby warm air rises and cold air falls, is important to the natural circulation and cleansing of the troposphere. Explain from a molecular point of view why convection occurs.
9. How does the kinetic molecular theory explain the fact that you can smell your mother's cooking the moment you enter your home?

EXTENDING YOUR KNOWLEDGE (OPTIONAL)

- In what ways is an "ocean of gases" analogy useful in thinking about the atmosphere? How does the atmosphere differ from the hydrosphere? (You may wish to compare a dive down into the ocean with a flight up through the atmosphere.)
- Boyle's law is illustrated each time you breathe in or breathe out. Explain.
- Boyle's law is a matter of life and death to scuba divers. On the surface of the water, the diver's lungs, tank, and body are under 1 atm pressure. However, under water the diver's body is under the combined pressure of atmosphere and water.
    a. When the diver is 34 ft below the surface of the ocean, how much pressure does his or her body experience?
    b. Why are pressurized tanks needed when diving to great depths?
    c. What would happen to the volume of the tank if it were not made strong enough to withstand such pressure?
    d. Why is it necessary to exhale and rise slowly when ascending from the ocean depths?
    e. How do the problems of a diver compare with those of a pilot climbing to a high altitude in an unpressurized plane?

- If air at 25 °C has a density of 1.28 g/L, and your classroom has a volume of 2×10^5 L,
  a. What is the mass of air in your classroom?
  b. If the temperature increased, how would this affect the mass of air in the room? Assume the room is not completely sealed.
- The kelvin temperature scale has played an important role in theoretical and applied chemistry.
  a. Find out whether zero kelvins (0 K) has been reached in the laboratory. If not, how close have scientists come to this temperature?
  b. At 0 K, how would matter behave?
  c. Research from the field of **cryogenics** (low-temperature chemistry and physics) has many possible applications. Do research in the library to learn about activities in this field.

The sulfur and nitrogen oxides generated here will probably be carried many miles across country to do their damage.

C ATMOSPHERE AND CLIMATE

Imagine a place where noonday sunshine on a rock makes it hot enough to fry an egg, where the night is cold enough to freeze carbon dioxide gas to dry ice, where the sun's ultraviolet rays are strong enough to burn your exposed skin in minutes. That place is the moon. Its extreme conditions are due to the absence of an atmosphere.

Together, the sun's radiant energy and the Earth's atmosphere help maintain a hospitable climate for life on our planet. The atmosphere delivers the oxygen gas we breathe and makes exhaled carbon dioxide available to plants for photosynthesis. It also serves as a sink for airborne wastes from people, industry, and technology—a role which causes increasing concern. And the atmosphere protects our skin from the sun's ultraviolet rays.

To understand the basis for our hospitable climate, it's important to know how solar energy interacts with the Earth's atmosphere. The sun warms the surface of the Earth. The Earth's warm surface, in turn, warms the air above it. Since warm air expands, its density decreases and it rises. Colder, more dense air falls. This gaseous movement creates continual air currents that drive the weather.

Let's look more closely at the sun's energy. What makes it so useful?

C.1 The Sunshine Story

As noted in the previous unit on nuclear energy, the enormous quantity of solar energy is generated by the fusion of hydrogen nuclei into helium. Most of this energy escapes from the sun as electromagnetic radiation. About 9% of solar radiation is in the **ultraviolet (UV)** region of the electromagnetic spectrum, 46% is in the **visible,** and 45% is in the **infrared (IR)** region of the spectrum. The complete solar spectrum is shown in Figure VI.11.

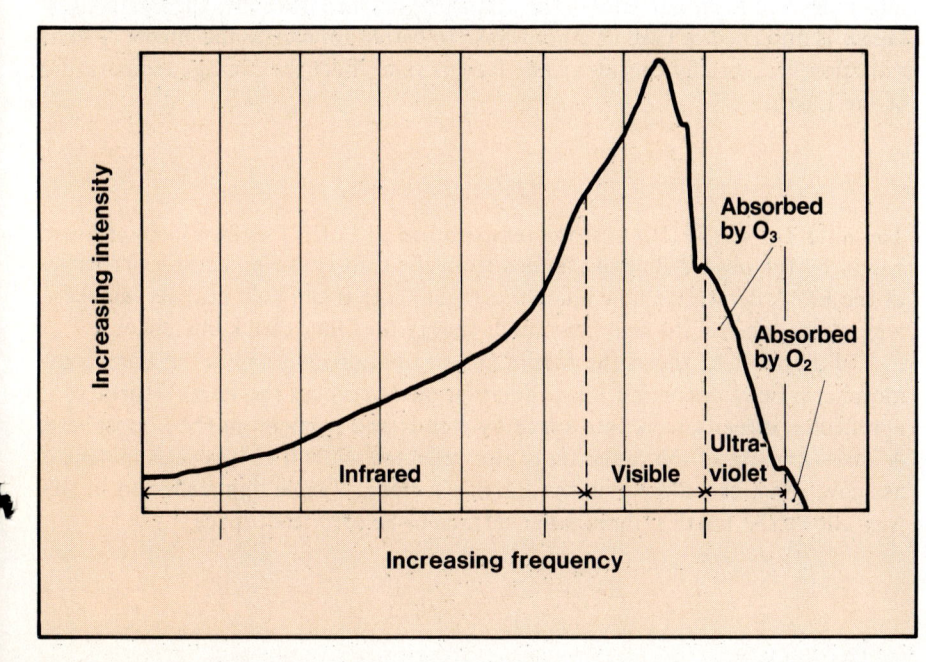

Figure VI.11 The solar spectrum. The higher the frequency, the higher the photon energy of the radiation. Intensity, plotted on the y axis, is a measure of the quantity of radiation at a given frequency.

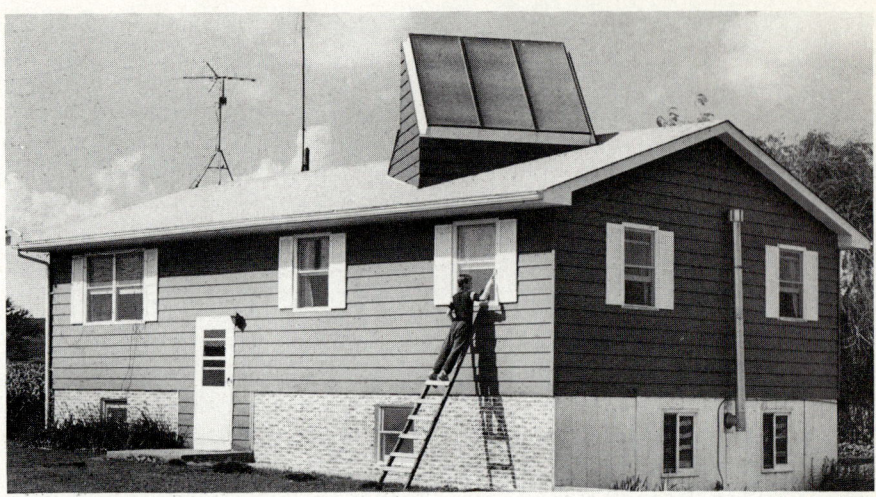

Utilizing the energy of the sun.

Electromagnetic radiation, as you learned in the previous unit, is composed of photons, each possessing a characteristic frequency and quantity of energy. The higher the frequency, the higher the energy of the photon. Photon energy determines the effect of radiation on matter.

Infrared radiation, with frequencies between 10^{12} and 10^{14} Hz (cycles/s), causes molecules to vibrate faster. As we have seen in our study of the kinetic molecular theory (Part B.8), this raises the temperature.

Visible light is of higher energy (frequency about 10^{14} Hz) and can energize electrons in some chemical bonds. One photon delivers its energy to the electrons in one bond. Such photon-electron interactions are taking place right now in double bonds of molecules in your eyes, making it possible for you to see this printed page. Visible light also interacts with electrons in chlorophyll molecules in green plants, providing the energy needed for the reactions of photosynthesis.

The still higher energy carried by photons of ultraviolet radiation, with frequencies ranging from 10^{14} to 10^{16} Hz, can break single covalent bonds. As a result, chemical changes can take place in materials exposed to ultraviolet radiation, including damage to tissues in living organisms.

As solar radiation passes through the Earth's atmosphere, it interacts with molecules and other particles in the atmosphere. Once it reaches the Earth's surface it is absorbed or used in other ways. To better appreciate the effects of solar radiation, let's briefly consider some factors that affect the average temperature of the Earth.

C.2 The Earth's Energy Balance

The mild 15 °C (59 °F) average temperature at Earth's surface is determined partly by the inward flow of photons from the sun. However, certain properties of the Earth determine how much thermal energy it can hold near its surface—where you can feel it—and how much energy it radiates back into space.

Figure VI.12 shows the fate of solar radiation as it enters the Earth's atmosphere. Some incoming solar radiation never reaches the Earth's surface. It is reflected directly back into space by clouds and particles in the atmosphere. A small amount of the radiation is also reflected when it strikes such materials as snow, sand, or concrete on the Earth's surface. Visible light reflected in this way allows the Earth's illuminated surface to be seen from space.

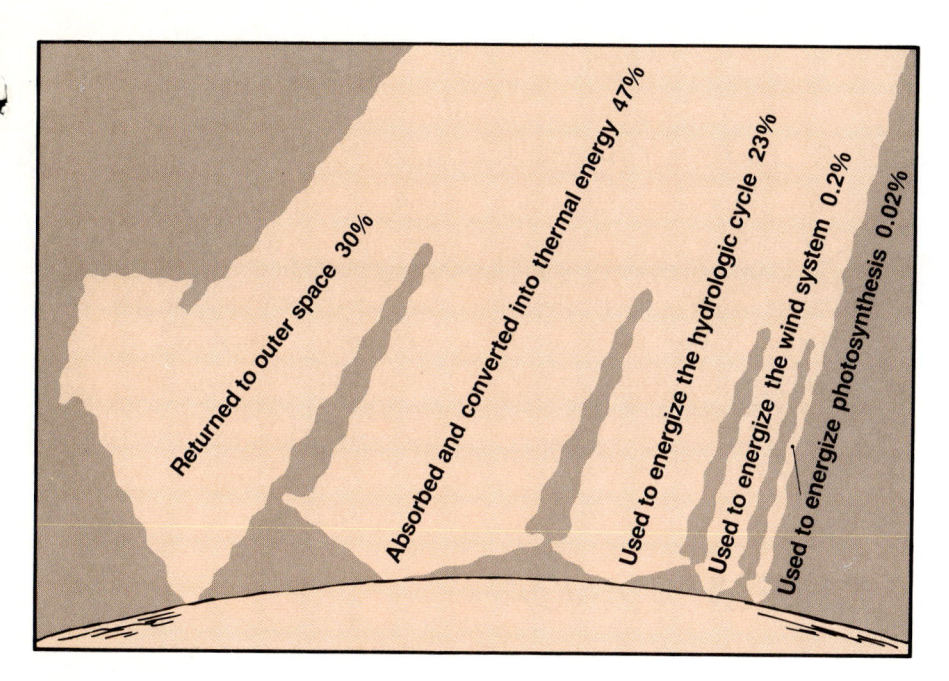

Figure VI.12 The fate of incoming solar radiation.

Returned to outer space 30%
Absorbed and converted into thermal energy 47%
Used to energize the hydrologic cycle 23%
Used to energize the wind system 0.2%
Used to energize photosynthesis 0.02%

About one-fourth of the incoming solar energy fuels the hydrologic cycle, which you learned in Unit One is the continuous cycling of water into and out of the atmosphere by evaporation and condensation.

Almost half of the solar energy is absorbed, warming the atmosphere, oceans, and continents. All objects above zero kelvins radiate energy, the quantity being directly related to the object's temperature. The Earth's surface re-radiates most of the absorbed radiation, not at the original frequencies, but at lower frequencies in the infrared region of the spectrum. This returning radiation plays an extremely important role in the Earth's energy balance. Its lower energy photons are absorbed by molecules in the air more easily than the original solar radiation, thus warming the atmosphere.

Carbon dioxide (CO_2) and water (H_2O) are good absorbers of infrared radiation. So are methane (CH_4) and halogenated hydrocarbons such as CF_3Cl. Clouds (concentrated droplets of water or ice) also absorb infrared radiation. The energy absorbed by these molecules in the atmosphere is once again re-radiated. Energy can pass back and forth between the Earth's surface and the molecules in the atmosphere many times before it escapes once again into outer space. This trapped energy keeps us warm.

Without water and carbon dioxide molecules in the atmosphere to absorb and re-radiate thermal energy to Earth, our planet would reach thermal balance at the lower temperature of −25 °C (−13 °F), quite close to the average temperature on Mars.

The trapping and returning of radiation by carbon dioxide and other substances in the atmosphere is referred to as the **greenhouse effect,** because it reminds people of the way heat is held in a greenhouse on a sunny day. The planet Venus provides an example of a runaway greenhouse effect. There the atmosphere is composed mainly of carbon dioxide and prevents the escape of most infrared radiation. Thermal balance is maintained at a much higher temperature than on Earth.

Check your understanding of the interaction of radiation with matter and the role of radiation in the Earth's energy balance by answering the following questions.

CO_2, H_2O, CH_4, CF_3Cl (and others) are "greenhouse gases."

YOUR TURN

VI.8 Earth's Energy Balance

1. Why is ultraviolet radiation potentially more harmful than infrared radiation?
2. Describe two essential roles played by visible radiation from the sun.
3. What might occur if human activities increase the concentration of carbon dioxide or other greenhouse gases in the troposphere?
4. Suppose Earth had a thinner atmosphere than it does at present.
   a. How would average daytime temperatures be affected? Why?
   b. How would average nighttime temperatures be affected? Why?

CHEMQUANDARY

Why does it cool off faster on a clear night than on a cloudy night?

Climate is influenced by other factors besides the interaction of solar radiation with the atmosphere. These factors include Earth's rotation (causing day and night and influencing wind patterns), its revolution around the sun (causing seasons), the uneven distribution of solar radiation across Earth's surface (influencing wind patterns), and the different thermal properties of materials on the Earth's surface. In the next section we will examine the last of these factors.

C.3 Thermal Properties at the Earth's Surface

If you have visited states in the South or Southwest (or if you live there), you may have noticed that many trucks and cars are light colored. A property of materials called **reflectivity** helps keep the vehicles cool. When light photons strike a surface some are absorbed, increasing the surface temperature, and some are reflected. The reflected radiation does not contribute to raising the object's temperature.

The color of an object is determined by the frequencies of the radiation it reflects. If all frequencies of the visible spectrum are reflected, it appears white. If all visible frequencies are absorbed (and none are reflected), the object appears black. Because light-colored surfaces reflect more radiation, they remain cooler than dark-colored surfaces.

Variations in the reflectivity of materials at the Earth's surface help to determine the surface temperature. On a hot day it is much more comfortable to walk barefoot across a plowed field than across an asphalt parking lot. The plowed field reflects almost 30% of the sun's rays, while the asphalt reflects very little. Clean snow reflects almost 95% of solar radiation, while forests are not very reflective.

CHEMQUANDARY

What might happen if large quantities of dust were to settle out on snow fields at the North and South Poles? What might happen if large quantities of dust of high reflectivity were to enter the atmosphere?

Each kind of material at the Earth's surface has a characteristic reflectivity and also a characteristic heat capacity, properties which together determine how fast the material warms up. The **heat capacity** is the quantity of thermal energy (heat) needed to raise the temperature of a given mass of a material by 1 °C. In effect, heat capacity is a measure of a material's thermal energy storage capability. The lower the heat capacity of a material, the greater will be its temperature increase for a given quantity of added energy. The higher the heat capacity, the smaller the temperature increase for a given quantity of added energy. Thus, materials with higher heat capacity can store more thermal energy.

Heat capacity is often given in units of joules per gram per °C (J/g °C).

Water is uniquely suited for its important role in the Earth's climate. One way is through its high heat capacity. Bodies of water are slow to heat up, to cool down, and can store large quantities of thermal energy. By contrast, land surfaces cool off much more rapidly and reach lower temperatures in the absence of sunlight. Oceans, lakes, and rivers, therefore, have a moderating effect on the temperature. For example, on a warm day breezes blow from the ocean toward the land, for the temperature over the land is higher. The warmer, less dense air rises, allowing cooler and more dense air from the ocean to move in. Then in the evening the land cools off more rapidly and the breezes shift direction, blowing from the now-cooler land toward the ocean. In general, temperatures fluctuate less near the ocean than in regions far from the coast.

YOUR TURN

VI.9 Thermal Properties of Materials

1. Which of the following would you expect to be cooler to the touch on a hot day in the sun:
   a. a concrete sidewalk or an asphalt sidewalk?
   b. a black plastic bicycle seat or a tan fur bicycle seat?
2. Would you expect the average temperature to be lower in a winter with large quantities of snow or with very little snow?
3. Why is water (heat capacity 4.2 J/g °C) a better fluid for a hot-water bottle than alcohol (heat capacity 2.6 J/g °C)?
4. Beach sand feels hotter than grass on a hot day. Which is more responsible for this observation—heat capacity or reflectivity?
5. Why does the temperature drop further on a clear, cool night than on a cloudy, cool night?
6. Which of the two medium-sized cities listed, at the same latitude and longitude, would you expect to be hotter in the summer? Why?
   a. A city with many asphalt roads and concrete buildings.
   b. The same city, but located near a large body of water.

Could the thermal balance of our planet be upset? Is human activity affecting climate? We'll examine these questions in the next section.

C.4 Drastic Changes on the Face of the Earth?

More than 100 years ago it was suggested that the rapidly increasing use of fossil fuels might release enough carbon dioxide into the atmosphere to change the climate. You read in Part C.2 that water and carbon dioxide in the atmosphere act as one-way screens. They let in sunlight, and then limit the escape of re-radiated infrared photons, producing the greenhouse effect, which keeps Earth at a comfortable average temperature.

Natural disasters such as the eruption of Mt. St. Helens can adversely affect atmospheric conditions.

The greenhouse effect is constant as long as the water vapor and carbon dioxide in the atmosphere remain at their normal concentrations and no significant amounts of other greenhouse gases are added. Both the hydrologic cycle (studied in Unit One) and the carbon cycle (Figure VI.13) maintain constant concentrations of water and carbon dioxide. However, human activity must be taken into consideration. In the case of water no problem is apparent. The atmosphere contains about 12 trillion metric tons of water vapor, a quantity so large that we cannot significantly affect it. There is much less carbon dioxide, however. Human activity has already increased its concentration by about 15% over the past century.

We increase CO_2 levels in several ways. In clearing forests we remove vegetation, which consumes CO_2 through photosynthesis (Figure VI.14). When limestone, a form of calcium carbonate, $CaCO_3$, is converted to concrete, some CO_2 is released. Most significantly, burning fossil fuels releases CO_2 into the air, as these equations illustrate:

$$\text{Burning coal: } C(s) + O_2(g) \rightarrow CO_2(g)$$
$$\text{Burning natural gas: } CH_4(g) + 2\ O_2(g) \rightarrow CO_2(g) + 2\ H_2O(g)$$
$$\text{Burning gasoline: } 2\ C_8H_{18}(l) + 25\ O_2(g) \rightarrow 16\ CO_2(g) + 18\ H_2O(g)$$

If more CO_2 is added to the atmosphere than can be removed by natural processes, its concentration can increase significantly. This could result in the atmosphere retaining enough additional infrared radiation to cause a rise in the Earth's surface temperature.

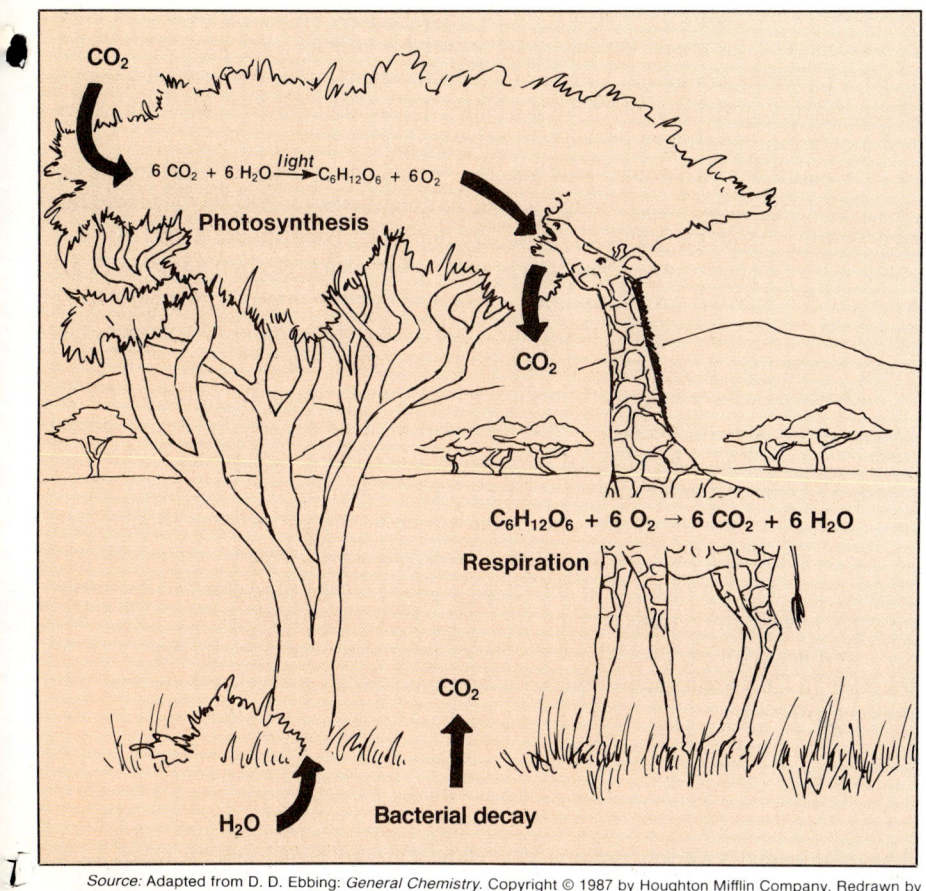

Figure VI.13 The carbon cycle. How does this figure illustrate the cycling of carbon throughout the biosphere?

Figure VI.14 Deforestation accounts for increases in atmospheric CO_2.

Will this occur? And if so, will the warming be enough to melt the polar caps, flood coastal cities, and change the Earth's climate? Perhaps the greatest impact in such a changing climate would be destruction of major food-growing areas. Some countries could lose all their arable land, while in other countries, the best grain-growing regions could change locations.

Scientists predict that doubling the preindustrial level of CO_2 in the atmosphere would increase average global temperatures by 2.8 °C. The greatest temperature increases would occur in the Northern Hemisphere above 40° latitude, where the burning of fossil fuels and seasonal changes in plant growth are greatest. Predictions are that above a line which runs through Northern California, Denver, Indianapolis, and Philadelphia, droughts would be common, and many fertile food-producing areas would turn to dust bowls.

Although increasing carbon dioxide will have a warming influence on the globe, no one can yet predict exactly what the climate will do. Too many factors affect climate, and most of these affect each other, often in ways not well understood. Human settlements warm the Earth by lowering its reflectivity—darkening it with cities and farms that replace forests and plains. Automobiles and air pollution affect local temperatures. Smog particles can both warm and cool the climate. On top of all this, the Earth's climate runs in cycles of alternating ice ages and warm periods.

Some scientists predict we are nearing another ice age. In that case, an increase in CO_2 might be just what the world needs to counteract that trend. But don't count on it.

C.5 *You Decide:* Examining Trends in CO_2 Levels

The CO_2 level data given in Table VI.3 show average measurements taken at the Mauna Loa Observatory in Hawaii and at the South Pole, where the air is well mixed, far from local CO_2 sources.

Part 1
1. Plot the data in Table VI.3. Prepare your horizontal axis to include the years 1870 to 2050. The vertical axis will represent CO_2 levels from 280 ppm to 600 ppm. Plot the appropriate points, and draw a smooth curve representing the trend of the plotted points. (Do not draw a straight line or attempt to connect every point. A smooth curve shows general trends.)
2. Assuming that the trend in your smooth curve continues from 1985 to 2050, extend your curve with a dashed line to the year 2050. This extrapolation is a prediction for the future based on past trends.

Part 2
You will now make some predictions using the graph you have just completed and evaluate these predictions. Keep in mind that extrapolations of this type are always tentative. Completely unforeseen factors may arise in the future.

1. What does your graph indicate about the general change in CO_2 levels since 1870?
2. Predict CO_2 levels today, in the year 2000, and in 2050, based on your extrapolation.
3. Does your graph predict a doubling of the 1870 CO_2 level?
4. Which predictions from Question 2 are the most accurate? Why?
5. What assumption do you make in extrapolating known data?
6. Describe factors that might cause these extrapolations to be incorrect.

Table VI.3 *Global air CO_2 levels*

Year	Approx. CO_2 level (ppm)
1870	291
1900	287
1920	303
1930	310
1960	317
1965	320
1970	325
1974	328
1976	330
1978	333
1980	337
1982	340
1983	342
1985	345

C.6 Laboratory Activity: Measuring CO₂ Levels

Getting Ready
The air we normally breathe has a very low CO_2 level. However, the CO_2 concentration in an area can be dramatically increased by burning coal or petroleum, decomposing organic matter, or accumulating a crowd of people or animals.

In this activity you will compare the amount of CO_2 in several air samples. To do this, the air will be bubbled through water that contains an indicator, bromthymol blue. Carbon dioxide reacts with water to form carbonic acid:

$$CO_2(g) + H_2O(l) \rightarrow H_2CO_3(aq)$$

As the concentration of carbonic acid in solution increases, bromthymol blue changes in color from blue to green to yellow.

Procedure

Part 1: CO_2 in Normal Air

1. Pour 125 mL of distilled or deionized water into a filter flask and add 10 drops of bromthymol blue. The solution should be blue. If it is not, ask your teacher for assistance in adjusting the acid content of the water.
2. Pour 10 mL of the solution prepared in Step 1 into a test tube. Put this aside; it is your control (to be used for comparison).
3. Assemble the apparatus illustrated in Figure VI.15 (without the candle).
4. Turn on the water tap so that the aspirator pulls air through the flask. Note the position of the faucet handle so you can run the aspirator at this same flow rate each time.
5. Let the aspirator run for 10 min. Turn off the water. Remove the stopper from the flask.
6. Pour 10 mL of the used indicator solution from the flask into a second test tube. Compare the color of this sample with the control. Save this test tube.

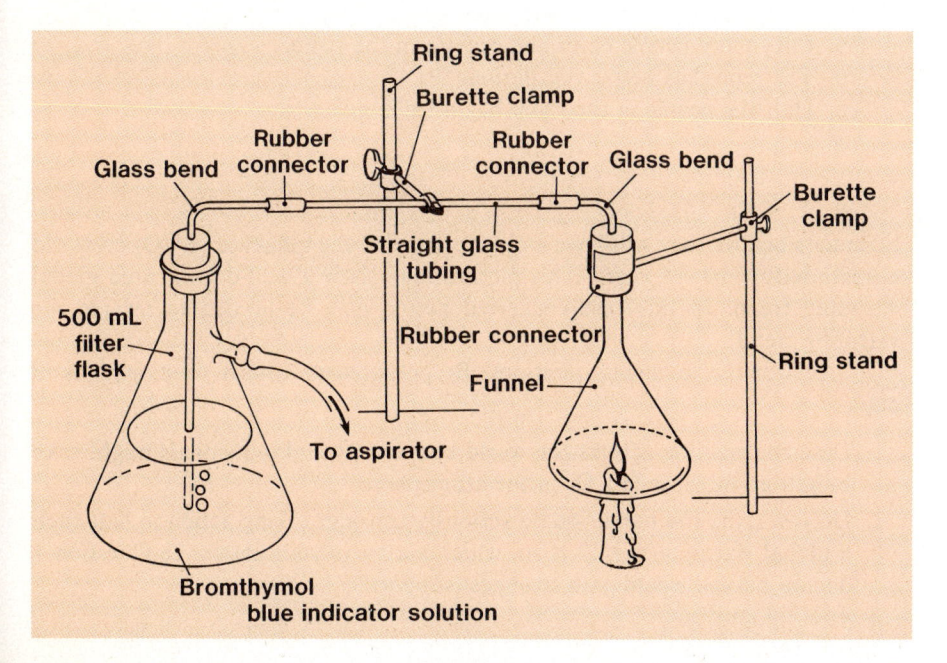

Figure VI.15 Apparatus for collecting air. Part 1 is done without candle.

Part 2: CO₂ from Combustion
1. Empty the filter flask and rinse it thoroughly with distilled or deionized water.
2. Refill the flask with indicator solution according to Step 1 in Part 1. Reassemble the apparatus as in Step 3 in Part 1.
3. Light a candle and position it and the funnel so that the tip of the flame is level with the funnel base.
4. Turn on the water tap to the preset mark. Note the starting time. Run the aspirator until the indicator solution turns yellow. Record the time this takes in minutes.
5. Pour 10 mL of the indicator solution into a clean test tube. Compare the color with that of the other two solutions. Record your observations.

Part 3: CO₂ in Exhaled Air
1. Pour 125 mL of distilled or deionized water into an Erlenmeyer flask and add 10 drops of bromthymol blue. As before, consult with your teacher if the color is not blue.
2. Note the time, then exhale into the solution through a clean straw, until the indicator color changes to yellow. Record the time this takes.
3. Pour 10 mL of the indicator solution into a clean test tube. Compare the color with that of your other three solutions. Record your observations.

Questions
1. Compare the colors of the indicator solutions from each test. Which sample contained more CO_2?
2. Compare the times and color changes with plain air and air from above the candle. Explain the differences.
3. Which contained more CO_2, air in the absence of the burning candle or exhaled air?
4. Indicate what effect the following would have on the color of the indicator solution:
   a. Plants were growing in the room.
   b. Fifty more people enter and remain in the room.
   c. The room has better ventilation.
   d. Half the people in the room begin smoking.

C.7 *You Decide:* Reversing the Trend

Scientists believe they will have a clear picture of the influence of CO_2 on world climate within 10–15 years. This picture may show a well-documented global warming trend. In this case concerted action may be needed to reverse the warming trend.

Answer the following questions. Be prepared to discuss your opinions in class.

1. Describe possible actions the world community and individuals might take to halt an increase in CO_2 in the atmosphere.
2. Do you think the public would seriously consider taking action to halt such a global warming trend? If so, which action in your answer to Question 1 do you believe would gain strongest support?

C.8 Off in the Ozone

Although a small dose of ultraviolet radiation is necessary for health, too much is dangerous. In fact, if all of the ultraviolet radiation in sunlight were to reach the Earth's surface, serious damage to life on Earth would occur. Ultraviolet photons, we have seen, have enough energy to break covalent bonds. The resulting chemical changes cause sunburn and cancer in humans and damage to many biological systems.

Fortunately, Earth has an ultraviolet shield high in the stratosphere. The shield consists of a layer of ozone, $O_3(g)$, which absorbs ultraviolet radiation.

As sunlight penetrates the stratosphere, the highest-energy ultraviolet photons react with oxygen molecules, splitting them into oxygen atoms. Individual oxygen atoms are very reactive. They immediately react further, most with oxygen molecules to produce ozone. A third molecule (typically N_2 or O_2, represented by M in the equation below) carries away excess energy but is unchanged.

Each ozone molecule formed can absorb a medium-energy ultraviolet photon in the stratosphere. Decomposition results, producing an oxygen molecule and an oxygen atom, which can then carry on the cycle by undergoing the second reaction to produce more ozone:

$$O_2 + \text{Ultraviolet photon} \rightarrow O + O$$
$$O_2 + O + M \rightarrow O_3 + M$$
$$O_3 + \text{Ultraviolet photon} \rightarrow O_2 + O$$

The concentration of ozone in the stratosphere is so small that if the molecules were located on the surface of the Earth at atmospheric pressure, they would surround the Earth in a layer only 3 mm thick (about the thickness of a hardback book cover).

Human activities may have endangered this fragile ozone layer. The major culprit may be chlorine atoms from chlorinated hydrocarbon molecules such as the Freons (e.g., CCl_3F). These substances have been used as propellants in aerosol cans and as cooling fluids in the sealed cooling systems of air conditioners and refrigerators.

Other gases that may deplete ozone: methane (CH_4) and other hydrocarbons; nitrogen monoxide (NO).

Although Freons are highly stable molecules, at an altitude of 30 km ultraviolet photons of suitable energy split chlorine atoms from them. The freed chlorine atoms can then participate in a series of reactions that destroy ozone:

$$Cl + O_3 \rightarrow ClO + O_2$$
$$ClO + O \rightarrow Cl + O_2$$

Notice that the chlorine atom consumed in the first reaction is regenerated in the second reaction. Thus, each chlorine atom can participate in a large number of these ozone-destroying reactions. The sum of these two reactions is

$$O + O_3 \rightarrow 2\, O_2$$

Chlorine atoms serve as catalysts here.

Freons are used in sealed refrigerator cooling systems.

The net effect of the reactions is the conversion of ozone molecules into oxygen molecules, triggered by free chlorine atoms released from molecules such as those of Freons.

Because of concern for the ozone layer, use of Freons in aerosol cans in the United States was banned in 1978. However, these substances are still widely used in this and other countries. In 1985 growing concern over depletion of the ozone layer, combined with the obvious need for international cooperation on this global problem, led 20 nations to sign an agreement to study means of protecting the ozone layer, possibly by limiting production and/or emissions.

We have seen that human activity may modify Earth's average temperature and its exposure to ultraviolet radiation by altering natural quantities of carbon dioxide and ozone. The chemistry of the atmosphere is very complex and difficult to study, and a complete picture of what is happening has not yet been developed. However, there is clear evidence of human alteration of the atmosphere in other ways. We'll look at this subject in the next few sections.

Other uses of Freons: Production of foam for insulation and packaging; sterilizing medical supplies.

PART C SUMMARY *QUESTIONS*

1. Compare infrared, visible, and ultraviolet radiation in terms of the relative energies of their photons. Cite one useful role each plays in a life process or a process important to life.
2. Explain why the Earth's atmosphere can be considered a one-way screen. How does this make Earth more hospitable to life?
3. List and briefly describe two functions of the stratosphere.
4. From a scientific viewpoint, why do tennis players and desert-dwellers wear white or light-colored clothing?
5. Compare ocean water and beach sand in terms of how quickly they heat up in the sun and how quickly they cool at night. What property of these two materials accounts for these differences?
6. Describe how atmospheric concentrations of CO_2 and water help to maintain moderate temperatures at the Earth's surface.
7. List the two ways humans increase the amount of CO_2 in the atmosphere. Which way involves greater contribution of CO_2?
8. What changes in atmospheric composition could result in:
    a. An increase in the average surface temperature of Earth.
    b. A decrease in the average surface temperature of Earth.
9. Describe an important role of the Earth's ozone shield.

EXTENDING *YOUR KNOWLEDGE* (OPTIONAL)

A satellite carrying instruments that monitor worldwide ozone concentration was launched in 1978. In 1986 many newspaper and magazine stories reported evidence from this satellite that an "ozone hole" was growing over Antarctica. The "hole" is a region where ozone depletion might be as high as 50%. Investigate the various theories proposed to explain the ozone hole.

D HUMAN IMPACT ON THE AIR WE BREATHE

Dirty air is so common in the United States that weather reports for some major cities now include indexes of certain air pollutants. Depending on where you live, cars, power plants, or industry may be the major polluter. However, air pollution is not only an outdoor problem. Indoor air is often highly polluted by people smoking or by fumes that spontaneously evaporate from certain products, such as from polymeric materials in furniture.

Air pollution smells bad, looks ugly, and blocks the view of the stars at night. But beyond being unpleasant, air pollution causes billions of dollars of damage every year. It corrodes buildings and machines. It stunts the growth of agricultural crops and weakens livestock. It causes or aggravates diseases such as bronchitis, asthma, emphysema, and lung cancer and so adds to the world's hospital bills. By some estimates, air pollution costs the United States $16 billion per year.

D.1 To Exist Is to Pollute

To threaten human health or damage property, the concentration of a pollutant must reach a harmful level in a specific location. Many natural processes emit substances that could become pollutants, but in most cases emission occurs over such a wide area that we do not notice it. Furthermore, the environment may dilute or transform substances before they can accumulate to harmful levels.

By contrast, pollution from human activities is usually generated in a small area. When the quantity of a pollutant overwhelms the ability of nature to dispose of it or disperse it, air pollution becomes a serious problem. Many large cities are prone to high concentrations of pollutants. If smog from Los Angeles were spread out over the entire West Coast, it would be much less noticeable.

Table VI.4 lists the major air pollutants and the quantities emitted worldwide from human and natural sources. These substances are all **primary air pollutants**—they enter the atmosphere in the chemical form listed. For example, methane (CH_4, the simplest hydrocarbon) is a by-product of fossil fuel use and a component of natural gas that leaks into the atmosphere. It is also produced by anaerobic bacteria and termites as they break down organic matter.

In addition to those listed in Table VI.4, there are several other important categories of air pollutants:

- *Secondary air pollutants.* These are substances formed in the atmosphere by chemical reactions between primary air pollutants and/or natural components of air. For example, sulfur dioxide (SO_2) reacts with oxygen in the air to form sulfur trioxide (SO_3), and the two oxides are always present together. Further reactions with water in the atmosphere can convert the sulfur oxides to sulfates (SO_4^{2-}) or sulfuric acid (H_2SO_4), a secondary pollutant partly responsible for acid rain (discussed in Part D.11).

SO_2 + SO_3 are referred to as SO_x ("socks").

Table VI.4 *Annual worldwide emissions of air pollutants (10^{12} g/yr or 10^6 metric tons/yr)*

Pollutant	Human source	Quantity	Natural source	Quantity
CO_2	Combustion of wood and fossil fuels	22,000	Biological decay; release from oceans, forest fires, and respiration	1,000,000
CO	Incomplete combustion (especially automotive)	700	Forest fires and photochemical reactions	2100
SO_2	Combustion of coal and oil; smelting of ores	212	Volcanoes and biological processes	20
CH_4	Combustion; natural gas leakage	160	Anaerobic biological decay and termites	1050
NO_x	High-temperature combustion (NO_x = NO and NO_2)	75	Lightning; bacterial action in soil	180
NMHC[a]	Incomplete combustion (especially automotive)	40	Biological processes	20,000
NH_3	Sewage treatment and fertilizers	6	Anaerobic biological decay	260
H_2S	Petroleum refining and sewage treatment	3	Volcanoes and anaerobic biological decay in soil and water	84

[a]NMHC = Nonmethane hydrocarbons.
Source: Adapted from Stern et al. *Fundamentals of Air Pollution,* 2nd ed.; Academic Press, Inc.: Orlando, FL, 1984; pp. 30–31. Table adapted by permission of Elmer Robinson, Mauna Loa Observatory.
Data based on conditions prior to 1980.

- *Particulates*. This major category of pollutants includes all solid particles that get into the air, either from human activities (e.g., power plants, waste burning, road building, mining) or natural processes (e.g., forest fires, wind erosion, volcanic eruptions). Particulates include visible emissions from smoke stacks or automobile tail pipes.
- *Manmade substances*. Some pollutants, such as the Freons, are produced only by human activity and would not otherwise be present at all.

Automobile emissions are a principal cause of air-quality problems in Denver.

D.2 *You Decide:* What Is Air Pollution? Where Does It Come from?

Use the information in Table VI.4 to answer the following questions.

1. What pattern was used to organize the data in Table VI.4? For what reason might this pattern have been chosen? Should the relative quantities of pollutants from human sources determine which pollutants should be reduced or controlled? If not, what other factors should be considered?
2. For all cases except one, natural contributions of these potentially polluting substances greatly exceed contributions from human activities.
    a. Does this imply that human contributions of these substances can be ignored? Why?
    b. For which substance does the contribution from human activity exceed that from natural sources? What does this suggest about modern society?
3. Refer to Table VI.1 in Part B.2. Which of the pollutants listed in Table VI.4 also occurs naturally in the atmosphere at concentrations of 0.0001% or more?
4. What is the major source of human contributions to air pollutants listed in Table VI.4?
5. Considering the large quantities of potential air pollutants from natural sources, why might adding small quantities of substances that do not exist naturally in the atmosphere pose a problem?

D.3 *You Decide:* Identifying Major Pollutants

Air pollution is a by-product of manufacturing, transportation, and energy production. Table VI.5 gives a detailed picture of the sources of some of the major air pollutants in the United States. Use this information to answer the following questions.

1. Overall, is industry the main source of air pollution? If not, what is the main source?
2. For which pollutants is one-third or more contributed by industry?

Table VI.5 *U.S. pollutant emissions by source, 1983 (in 10^6 metric tons/yr)*

	Pollutant (See symbol explanations below)					
Source	TSP	SO_x	NO_x	HC	CO	Total
Transportation (petroleum burning)	1.3	0.9	8.8	7.2	47.7	65.9
Fuel burning for space heating and electricity	2.0	16.8	9.7	—	—	28.5
Industrial processes	2.3	3.1	—	7.5	4.6	17.5
Solid waste disposal and miscellaneous	1.3	—	0.9[a]	5.2[b]	15.3	22.7
Totals	6.9	20.8	19.4	19.9	67.6	134.6

Symbols used:
TSP = Total suspended particulates.
SO_x = Sulfur oxides (SO_2 and SO_3).
NO_x = Nitrogen oxides (NO and NO_2).
HC = Volatile organic compounds (methane and nonmethane hydrocarbons).
CO = Carbon monoxide.
[a]This value is a composite of industrial processes and the solid waste, miscellaneous category.
[b]This value includes a small contribution from fuel consumption and nonindustrial solvent use.

3. For which pollutants is one-third or more contributed by transportation?
4. For which pollutants is one-third or more contributed by burning fuel for space heating and electricity (usually referred to as "stationary fuel burning")?

How much pollution do *you* produce? If you drive, you are a major polluter—automobiles contribute about half the total mass of air pollutants. When you spend time in heated buildings, use electricity, or buy food and other products, air has been polluted for your benefit. As a comic strip character once said, "We have met the enemy, and he is us."

D.4 Smog: Spoiler of Our Cities

When weather forecasters give the air quality index along with humidity and temperature, to what are they referring? The U.S. Environmental Protection Agency has devised an index based on concentrations of pollutants that are major contributors to smog over cities. You can see in Table VI.6 that the combined health effects of these pollutants can become quite serious. Smog can kill. During one of the deadliest smogs, in London in 1952, the death rate among residents was more than double the normal rate. Similar although less severe episodes have occurred in other cities before and since.

The composition of smog depends on the type of industrial activity and power generation in an area and also on climate and geography. Over large cities containing many coal- and oil-burning industries, power-generating stations, and homes, the principal components of smog are sulfur oxides and particulates.

Coal and petroleum both contain varying quantities of sulfur, from which sulfur oxides are formed during combustion. One successful approach to improving air quality sets limits on the quantity of sulfur that may be present in the fuels burned.

Table VI.6 *U.S. Pollutant Standards Index (PSI)*[a]

		Air pollutant levels (micrograms per cubic meter)					
Air quality index value	Air quality description	Total suspended particulate matter (24 hours)	Sulfur dioxide (24 hours)	Carbon monoxide (8 hours)	Ozone (1 hour)	Nitrogen dioxide (1 hour)	Effects and suggested actions
500	Hazardous	1,000	2,620	57,500	1,200	3,750	Healthy people experience symptoms that affect normal activity. All should remain indoors with windows and doors closed, minimize physical exertion and avoid vehicular traffic. Premature death of some in high risk group.[a]
400	Hazardous	875	2,100	46,000	1,000	3,000	High risk individuals should stay indoors and *avoid* physical activity. Decreased exercise tolerance in healthy persons. General population should avoid outdoor activity.
300	Very unhealthful	625	1,600	34,000	800	2,260	High risk group has significant aggravation of symptoms and decreased exercise tolerance and should stay indoors and *reduce* physical activity. Widespread irritation symptoms in the healthy population.
200	Unhealthful	375	800	17,000	400	1,130	Those with lung or heart disease should reduce physical exertion and outdoor activity. Healthy individuals notice irritations.
100	Moderate	260	365	10,000	235	Not reported	Some damage to materials and vegetation but human health not affected unless levels continue for many days.
50	Good	75	80	5,000	118	Not reported	No significant effects.
0		0	0	0	0	0	

[a]High risk group includes elderly people and those with heart or lung diseases.
Source: U.S. Environmental Protection Agency.

Particulates from burning fossil fuels consist of unburned carbon or solid hydrocarbon fragments and trace minerals. Some particles contain toxic compounds of metals such as cadmium, chromium, lead, or mercury.

The fatality rate in severe smogs has been higher than predicted from known hazards of sulfur oxides or particulates alone. According to some researchers, this may be due to **synergistic interactions** in which the combined effect of the two pollutants is greater than the sum of the effects of either one alone. One way this can happen is that smaller particles have more surface area per unit mass than do larger particles, allowing them to absorb and concentrate greater amounts of gaseous pollutants. Also, the smaller the particles, the more likely they are to be carried into the lungs and bloodstream.

Before discussing smog resulting from automobile emissions, we will look at what can be done to decrease pollution, and what industry is doing to clean up its smoke.

A synergistic interaction results when a total effect is greater than the sum of the individual contributing factors.

D.5 Pollution Control

There are several basic ways to control air pollution:

- Energy technologies that cause air pollution can be replaced with technologies that don't require combustion, such as solar power, wind power, and nuclear power.
- Pollution from combustion can be reduced by energy conservation measures, such as getting more from what we burn and therefore burning less.
- Pollution-causing substances can be removed from fuel before burning. For example, most sulfur can be removed from coal.
- The combustion process can be modified so that fuel is more completely oxidized.
- Pollutants can be trapped after combustion.

All pollution control options cost money. When deciding upon a pollution control strategy, two key questions must be answered: What will the pollution control cost? What benefits will the pollution control offer?

CHEM*QUANDARY*

Consider this pollution control option: "Clean the atmosphere after the pollution is emitted." Is this a reasonable or practical strategy? Explain your answer. Would cloud-seeding to cause rain, or issuing filtering devices such as gas masks to individuals be reasonable alternatives? Explain.

D.6 Controlling Industrial Emission of Particulates

Power plants and smelters generate more than 60% of the particulate matter emitted in the United States. Several cost-effective methods for controlling particle emissions are being used; in recent years great progress has been made in cleaning up this type of pollution.

Electrostatic precipitation. This is currently the most important technique for controlling pollution by particulates. Combustion by-products pass through a strong electrical field, where they become charged. The charged particles collect on plates of opposite charge. This technique removes up to 99% of the total particulates, leaving only particles smaller than one-tenth of a micrometer (0.1 μm, where 1 $\mu m = 10^{-6}$ m). Dust and pollen collectors installed in home ventilation systems are often based on this technique.

Mechanical filtering. This works like the bag in a vacuum cleaner. Combustion by-products pass through a cleaning room (bag house) where huge filters trap up to 99% of the particles.

Cyclone collection. Combustion by-products pass rapidly through a tight circular spiral chamber. Particles thrown outward to the walls drop to the base of the chamber where they are removed. This technique removes 50–90% of larger, visible particles but relatively few of the more harmful particles (those smaller than 1 µm).

Scrubbing. This method controls particles and the sulfur oxides accompanying them. Substances that react with the pollutants are added to the stream of combustion by-products.

In dry scrubbing (on the left in Figure VI.16) powdered limestone (calcium carbonate, $CaCO_3$) is blown into the combustion chamber where it decomposes:

$$CaCO_3(s) + \text{Heat} \rightarrow CaO(s) + CO_2(g)$$

The lime (CaO) then reacts with sulfur oxides to form calcium sulfite ($CaSO_3$) and calcium sulfate ($CaSO_4$):

$$CaO(s) + SO_2(g) \rightarrow CaSO_3(s)$$
$$CaO(s) + SO_3(g) \rightarrow CaSO_4(s)$$

These products are washed away as a slurry (a mixture of solids and water).

Wet scrubbing (on the right in Figure VI.16) removes sulfur dioxide using an aqueous solution of calcium hydroxide, $Ca(OH)_2(aq)$ produced by the reaction of lime (CaO) with water. The reactions are as follows:

$$CaO(s) + H_2O(l) \rightarrow Ca(OH)_2(aq)$$
$$Ca(OH)_2(aq) + SO_2(g) \rightarrow CaSO_3(s) + H_2O(l)$$

Scrubbers can remove up to 95% of sulfur oxides. They are required for all new coal-burning plants in the United States. Their use adds significantly to the cost of electrical power.

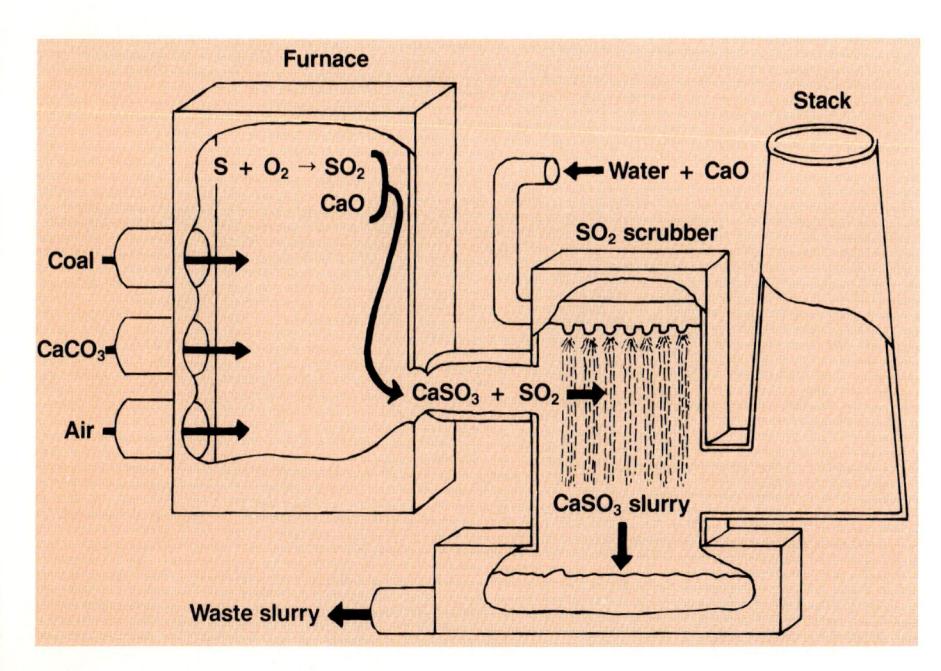

Figure VI.16 A scrubber for removing sulfur dioxide and particulate matter from products of industrial combustion processes. Dry scrubbing occurs in the furnace and wet scrubbing in the SO_2 scrubber.

D.7 Laboratory Demonstration: How Industry Cleans Its Air

Your teacher will demonstrate two pollution control methods.

Part 1: The Electrostatic Precipitator
1. Observe what happens to the smoke. Record your observations.
2. Observe the chemical reaction that occurs on the copper rod. Record your observations.

Part 2: Wet Scrubbing
1. Observe the color of the liquid and pH paper in each flask as the reaction proceeds. Record your observations.

Questions
1. Write an equation for the reaction that occurs between HCl and the NH_3 in Part 1.
2. What information does the universal indicator provide in each flask in Part 2?
3. What information does the pH paper at the neck of each flask provide?
4. What is the overall effect of the scrubbing, as shown by the indicators?
5. List the two ways the quality of the air in the reaction vessel was changed by wet scrubbing.
6. What advantages do precipitators have over wet scrubbers? What are their disadvantages?

CHEM*QUANDARY*

Explain why treating smoke before it is released from power plants is an important goal, but why, at the same time, it may mislead the general public concerning what is required to obtain clean air.

YOUR*TURN*

VI.10 Pollutants by the Ton

Large coal- or oil-fired power plants and large smelters can produce tons of pollutants or tons of scrubbing products in each hour of operation. You can estimate the amount produced from such operations using what you learned earlier about mass and molar mass relationships (Your Turn VI.1).

Assume that a coal-fired power plant burns 1.0×10^6 kg (or 1000 metric tons) of coal each hour. The coal contains 3.0% sulfur by mass. If the sulfur were converted to $SO_2(g)$ during combustion, how many moles of $SO_2(g)$ would be released to the atmosphere each hour? How many tons of $SO_2(g)$ is this?

First, we can find the mass in kilograms of sulfur in the coal burned each hour:

$$\left(\frac{1.0 \times 10^6 \text{ kg coal}}{1 \text{ h}}\right) \times \left(\frac{3.0 \text{ kg S}}{100 \text{ kg coal}}\right) = 3.0 \times 10^4 \text{ kg S/h}$$

1 metric ton = 1000 kg = 1×10^6 g

From this we can calculate the moles of sulfur produced each hour:

$$\left(\frac{3.0 \times 10^4 \text{ kg S}}{1 \text{ h}}\right) \times \left(\frac{1000 \text{ g S}}{1 \text{ kg S}}\right) \times \left(\frac{1 \text{ mol S}}{32 \text{ g S}}\right) = 9.4 \times 10^5 \text{ mol S/h}$$

Knowing that sulfur burns to form SO_2 according to the equation

$$S(s) + O_2(g) \rightarrow SO_2(g)$$

and observing that each mole of sulfur produces one mole of SO_2, we recognize that 9.4×10^5 mol of SO_2 will be released each hour. The mass in metric tons of this quantity of SO_2 is

$$\left(\frac{9.4 \times 10^5 \text{ mol } SO_2}{1 \text{ h}}\right) \times \left(\frac{64 \text{ g } SO_2}{1 \text{ mol } SO_2}\right) \times \left(\frac{1 \text{ metric ton}}{1 \times 10^6 \text{ g } SO_2}\right)$$

$$= 60 \text{ metric tons } SO_2/\text{h}$$

Now it's your turn.

1. If the exhaust gases in the sample problem above are dry-scrubbed with lime (CaO) according to the reaction shown in Part D.6, what mass of calcium sulfite, $CaSO_3(s)$, would be formed each hour?
2. A coal-producing company is considering removing sulfur from their coal and using the sulfur to produce sulfuric acid (H_2SO_4). The sulfuric acid can be sold at a small profit. The coal contains 7.5% sulfur (by mass). Forty percent of the sulfur can be removed by a known method. (a) How many moles of sulfur can be removed from each 5.0 kg of coal? (b) How many kilograms of sulfuric acid can be produced from this sulfur? (The molar mass of sulfuric acid is 98 g/mol.)

Furnaces and power plants are not the only sources of smog. Automobiles contribute many air pollutants.

D.8 Photochemical Smog

The ill effects of pollution from automobiles were first noted in the Los Angeles area in the 1940s. A brownish haze that irritated the eyes, nose, and throat and damaged crops appeared in the air. It puzzled researchers for some time. Los Angeles has no significant industrial or heating activities. The city has an abundance of automobiles and sunshine. Also, mountains rise on three sides of it. These geographic conditions produce temperature inversions about 320 days each year.

Normally, air at the Earth's surface is warmed by solar radiation and by re-radiation from surface materials. This warmer, less dense air rises, carrying pollutants with it. Cooler, less polluted air then moves in. In a temperature inversion, a cool air mass is trapped beneath a less dense warm air mass, often in a valley or over a city. Pollutants cannot escape and their concentration may rise to dangerous levels. Los Angeles is smog-prone, primarily because of its location. But there is more to the story for Los Angeles smog was much worse than seemed reasonable.

As so often happens in science, a serendipitous discovery provided a piece of the puzzle of smog. In 1952, chemist Arie J. Haagen-Smit was attempting to isolate the main ingredient of pineapple odor. Working on a smoggy day, he detected a greater concentration of ozone (O_3) in his experiment than is normally found in clean, tropospheric air. He delayed his research to identify its source. Within a year, he published a ground-breaking paper, "The Chemistry and Physics of Los Angeles Smog," in which he described the importance of sunlight in smog chemistry, and coined the term **photochemical smog**.

Any reaction initiated by light is a photochemical reaction.

Nitrogen oxides are essential ingredients of such smog. At the high temperature and pressure of automotive combustion (2800 °C and about 10 atm), nitrogen and oxygen react to produce the pollutant nitrogen monoxide (NO).

$$N_2(g) + O_2(g) + \text{Energy} \rightarrow 2\ NO(g)$$

In the atmosphere, nitrogen monoxide is oxidized to orange-brown nitrogen dioxide (NO₂).

$$2\ NO(g) + O_2(g) \rightarrow 2\ NO_2(g)$$

Carbon monoxide from automobile exhaust is also present.

The photochemical smog cycle begins as photons from sunlight initiate dissociation of NO₂ into NO and oxygen atoms (O). The atomic oxygen then reacts with oxygen molecules to produce ozone in the same way that it does in the stratosphere.

$$NO_2(g) + \text{Energy} \rightarrow NO(g) + O(g)$$
$$O_2(g) + O(g) \rightarrow O_3(g)$$

We have now accounted for two of the harmful and unpleasant ingredients of photochemical smog: Nitrogen dioxide has a pungent, irritating odor. Ozone is a very powerful oxidant. At concentrations as low as 0.1 ppm ozone can crack rubber, corrode metals, and damage plant and animal tissues.

The highly reactive ozone undergoes a complex series of reactions with the third essential ingredient of photochemical smog—hydrocarbons that escape from gasoline tanks or are emitted during incomplete combustion of gasoline. The products of these reactions cause burning eyes, are harmful to individuals with respiratory or heart disease, and can injure plants and damage materials such as rubber and paint.

For our purposes, the following equation represents the key ingredients and products of photochemical smog:

NO + NO₂ are referred to as NOₓ ("nocks").

$$\text{Auto exhaust} + \text{Sunlight} + O_2(g) \rightarrow$$
$$(\text{Hydrocarbons} + CO + NO_x)$$

$$O_3(g) + NO_x(g) + \text{Organic compounds} + CO_2(g) + H_2O(g)$$
$$\text{(oxidants and irritants)}$$

Table VI.7 shows how the concentrations of some important components of photochemical smog can vary from a relatively clear day to a smoggy day. Complete the following activity to discover how some of the key reactions in photochemical smog were pieced together.

Table VI.7 *Varying concentrations of smog components[a]*

	Concentration (ppm)		
Pollutant	Clear day	Smoggy day	Increase
Carbon monoxide	3.5	23.0	×6.5
Hydrocarbons	0.2	1.1	×5.5
Peroxides	0.1	0.5	×5.0
Oxides of nitrogen	0.08	0.4	×5.0
Lower aldehydes	0.07	0.4	×6.0
Ozone	0.06	0.3	×5.0
Sulfur dioxide	0.05	0.3	×6.0

[a]Clear day visibility is 7 mi; smoggy day visibility is 1 mi.
Source: From *Chemistry, Man and Society,* 4th ed. by Jones, M. M.; Johnston, D. O.; Netterville, J. T.; Wood, J. L. © 1983 by CBS College Publishing. © 1984, by Saunders College—Holt, Rinehart, & Winston. Copyright 1972 and 1976 by W. B. Saunders Co. Reprinted by permission of Holt, Rinehart & Winston, Inc.

D.9 *You Decide:* Automobile Contributions to Smog

Use the data in Figure VI.17 to answer these questions.

1. Between what hours do the concentrations of nitrogen oxides and hydrocarbons peak? Account for this fact in terms of traffic patterns.
2. Give two reasons why a given pollutant may be observed to decrease in concentration over several hours.
3. The concentration maximum for NO_2 occurs at the same time as the concentration minimum of NO. Explain this phenomenon.
4. Although ozone is necessary in the stratosphere to protect us from ultraviolet light, on the surface of the Earth it is a major component of photochemical smog. Determine from Figure VI.17 which chemicals, or species, are at minimum concentrations when $O_3(g)$ is at maximum concentration. What does this suggest about the production of $O_3(g)$ in polluted tropospheric air?

Smog is being produced faster than the atmosphere can dispose of it. However, the nation has made considerable progress in smog control. Many cities have cleaner air than they did 30 years ago. The following section will explore the control of pollution from automobiles.

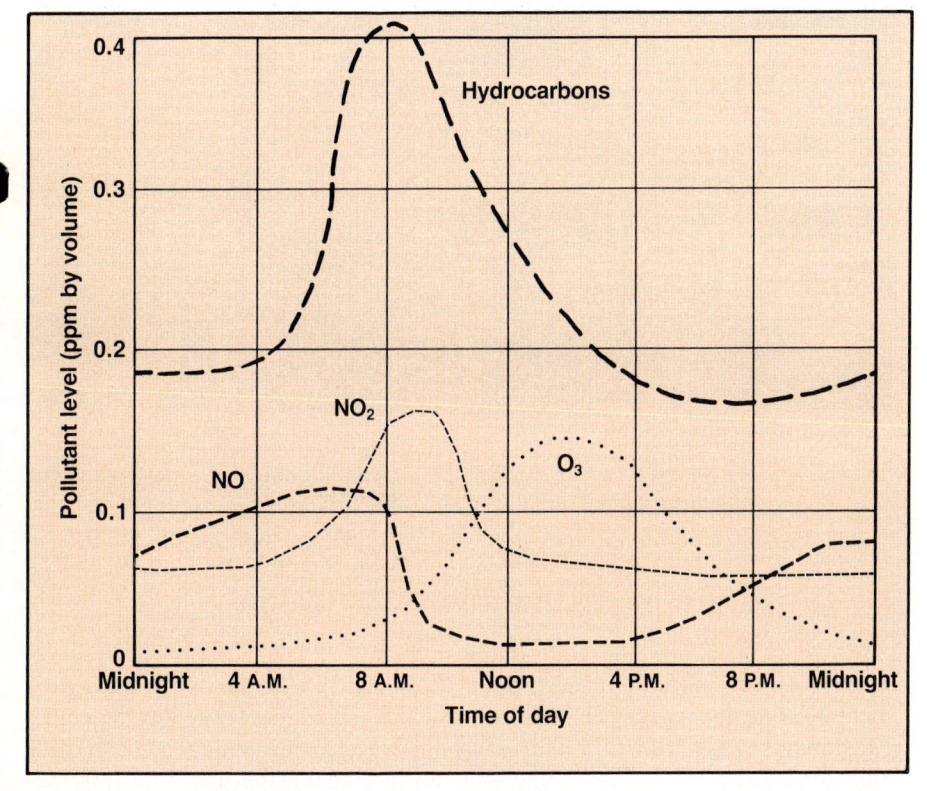

Figure VI.17 Species involved in photochemical smog formation.

Adapted from Manahan, Stanley, "General Applied Chemistry," 2nd ed. William Grant Press, Boston, 1982, p. 282.

Chemical & Engineering News

Exhaust emissions from slow-moving traffic on crowded freeways is a major source of the Los Angeles photochemical smog shown in this 1970 photo.

D.10 Controlling Automobile Emission of Pollutants

In 1970 an amendment to the Federal Clean Air Act authorized the Environmental Protection Agency (EPA) to set emissions standards for new automobiles. Maximum limits were set for hydrocarbon, nitrogen oxide, and carbon monoxide emissions. These values were to be achieved in gradual steps by 1975. Improvements in automobile engines were made between 1970 and 1975 by modifying the air-fuel ratio, adjusting the spark timing, adding carbon canisters to absorb gasoline that would normally evaporate before combustion, and installing exhaust gas recirculation systems.

To decrease emissions enough to meet the standards completely, however, required further measures. The result was development of the **catalytic converter.** The converter is a reaction chamber attached to the exhaust pipe. The exhaust gases and outside air pass over several catalysts which help convert nitrogen oxides to molecular nitrogen and hydrocarbons to carbon dioxide and water. The

Emissions testing has become an essential component of tuning an automobile.

carburetor air-fuel ratio is set to produce exhaust gases with relatively high concentrations of carbon monoxide and hydrogen. These gases enter the first half of the catalytic converter where nitrogen oxides are reduced, for example,

$$2\ NO(g) + 2\ CO(g) \xrightarrow{\text{Catalyst}} N_2(g) + 2\ CO_2(g)$$

$$2\ NO(g) + 2\ H_2(g) \xrightarrow{\text{Catalyst}} N_2(g) + 2\ H_2O(g)$$

The second half of the converter then further oxidizes carbon monoxide and hydrocarbons to carbon dioxide and water.

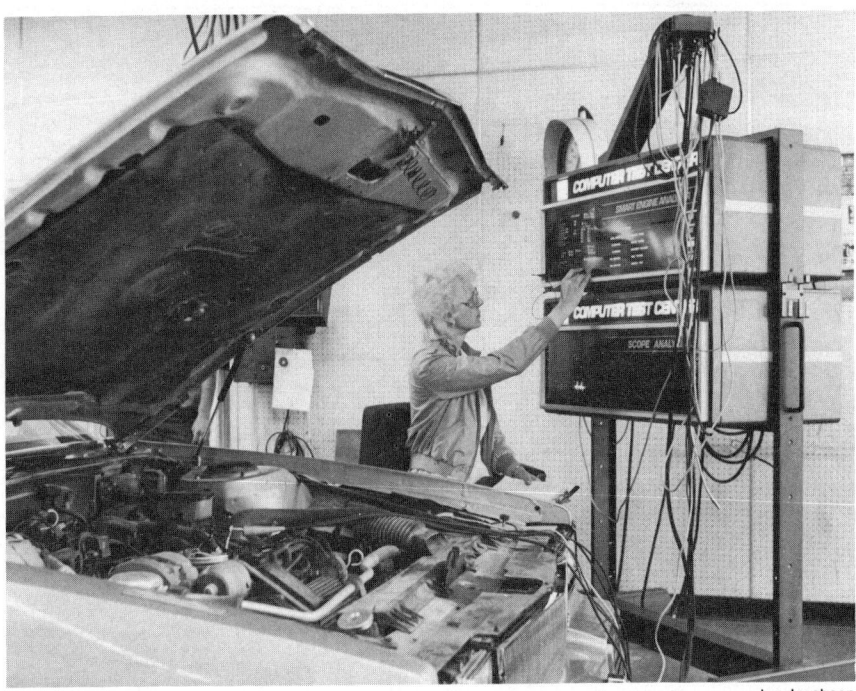

What exactly is a catalyst? You have encountered catalysts several times now. Enzymes that aid in digesting food and in other body functions are biochemical catalysts. The manganese dioxide added in generating oxygen in the laboratory activity in Part B.1 was a catalyst. In every case, the catalyst increases the rate of a chemical reaction which, without the catalyst, would proceed too slowly to be useful. A catalyst is not considered a reactant because it remains unchanged after the reaction is over.

How can a catalyst speed up a reaction and escape unchanged? Reactions between two or more species can occur only if their molecules collide with sufficient energy and the correct orientation to disrupt bonds. The minimum energy required for such effective collisions is called the **activation energy.** You can think of the activation energy as an energy barrier that stands between the reactants and products. Reactants must have enough energy to mount the barrier before reaction can occur. The higher the barrier, the fewer the molecules that have the energy to mount it, and the slower the reaction proceeds.

A catalyst increases the rate of a chemical reaction by providing a different reaction pathway, one with a lower activation energy. The result is that more molecules have sufficient energy to react within a given period of time. Thus, more product is formed. In automotive catalytic converters, 1–3 g of such metals as platinum, palladium, and rhodium act as catalysts.

Table VI.8 *Passenger car regulatory exhaust emission control requirements in the United States (emissions are expressed in grams per mile [g/mi])*

Model year	HC	CO	NO$_x$
1975–76	1.5	15	3.1
1977–79	1.5	15	2.0
1980	0.41	7.0	2.0
1981–82	0.41	3.4[a]	1.0[b]
1983	0.41	3.4	1.0
Pre-1968 (uncontrolled) values	8.6	87.5	3.5

[a]Possible two-year waiver to 7 g/mi.
[b]Possible waiver to 1.5 g/mi for diesel or innovative technology through 1984.
Source: Taylor, K. C. "Automobile Catalytic Converters," in *Catalysis Science and Technology;* Anderson, J. R., Boudart, M., Eds.; Springer-Verlag: New York, 1984; V. 5, Ch. 2, pp. 120–70.

Table VI.8 shows the gradually decreasing emission limits over the years called for by legislation. While catalytic converters have contributed to cleaning up automobile exhaust, the legislated goals have not yet been met. In particular, it has proved difficult to meet the nitrogen oxides standards.

CHEM*QUANDARY*

From a practical point of view, why might controlling air pollution from our nation's 140 million automobiles be more difficult than controlling air pollution from our power plants and industries?

D.11 Acid Rain

It started in Scandinavia. Then, it appeared in the northeastern United States. Now, the problem seems to exist in much of the industrialized world. Fish disappear from major lakes. The surfaces of limestone and concrete buildings and marble statues crumble. Crops grow more slowly and forests begin to die out. Although New England and Scandinavia are among the regions hardest hit by acid rain, it is widespread in the industrialized world. Even the Grand Canyon suffers from acid rain due to air pollution from coal-fired power plants miles away.

Although the cause of German forest die-outs is subject to debate, the rest of these problems have been traced definitively to acid rain. Naturally occurring substances, chiefly carbon dioxide, have always caused rainwater to be slightly acidic. Normally, rainwater's pH is about 5.6. Carbon dioxide reacts with rainwater to form carbonic acid:

$$CO_2(g) + H_2O(l) \rightarrow H_2CO_3(aq)$$

Field Museum of Natural History (neg. #PP-173-a), Chicago

Air pollutants are responsible for the partial disintegration of this classic Greek sculpture.

Oxides of sulfur and nitrogen emitted from power plants, various industries, and automobiles react with rainwater to form acids that have lowered the pH of rainwater to 4–4.5 in the northeastern United States. The key reactions are

$$H_2O(l) + SO_2(g) \rightarrow H_2SO_3(aq)$$
Sulfurous acid

$$H_2O(l) + SO_3(g) \rightarrow H_2SO_4(aq)$$
Sulfuric acid

$$H_2O(l) + 2\ NO_2(g) \rightarrow HNO_3(aq) + HNO_2(aq)$$
Nitric acid   Nitrous acid

Occasionally the levels of these oxides in air produce enough acid to lower the pH to 3. (A pH of 4–4.5 is about the acidity of orange juice; a pH of 3 is about that of vinegar.)

The lower the pH, the more acidic the solution.

Sulfuric acid contributes close to two-thirds of the acidity of rain in New England; acids of nitrogen contribute most of the other third.

This more acidic rain lowers the pH of lakes, killing fish eggs and other aquatic life; some species are more sensitive than others. Statues and monuments, such as the Parthenon in Greece, which have stood uneroded for centuries, suddenly began corroding because of acid rain and sulfates. The acid attacks calcium carbonate in limestone, marble, and concrete:

$$H_2SO_4(aq) + CaCO_3(s) \rightarrow CaSO_4(s) + H_2O(l) + CO_2(g)$$

Calcium sulfate is more soluble than calcium carbonate. Therefore, the stonework decays as the calcium sulfate is washed away to uncover fresh calcium carbonate that can react further with the acid rain.

Salts of sulfuric acid, which contain the sulfate ion (SO_4^{2-}), are also present in acid rain and atmospheric moisture. In the air they scatter light, causing haze. And they are potentially harmful to human health if they are deposited in large enough amounts in the lungs.

Many different and interrelated reactions are involved in producing acid rain. Studies aimed at a more complete understanding of the causes are under way. One puzzle is how sulfur dioxide is oxidized to sulfur trioxide. Oxygen dissolved in water oxidizes sulfur dioxide very slowly. The reaction may be accelerated in the atmosphere by sunlight or by catalysts such as iron, manganese, or vanadium in soot particles.

The control of acid rain is made difficult because air pollution knows no political boundaries. Acid rain often appears hundreds of kilometers from the sources of pollutants. For example, much of the acid that rains on Scandinavia is thought to come from Germany and the United Kingdom. The rain that falls on New England may come largely from the industrial Ohio Valley, in the Midwest.

Next, you will have an opportunity to work directly with solutions having acidity approximately equal to that of acid rain.

D.12 Laboratory Activity: Acid Rain

Getting Ready
In this activity you will create a mixture similar to that in acid rain by burning sulfur in air and then adding water. You will observe the effects of acid rain chemistry on plant material (represented by an apple peel), living creatures (represented by a culture of microorganisms), an active metal, and marble.

Procedure

1. Cut a piece of skin from an apple, and place it in an empty 500-mL glass bottle.
2. Fill a combustion spoon half full of powdered sulfur.
3. *In a fume hood:* Turn on a tap water spigot so that it drips slowly. (If there is no spigot in the hood, adjust the one at your laboratory bench, and return to the fume hood to perform the following procedures.) Light the Bunsen burner. Ignite the sulfur by holding the combustion spoon over the burner flame until you observe a blue flame. Quickly insert the spoon into the bottle and cover as much of the bottle opening as possible with a glass plate.
4. When smoke fills the bottle, remove the spoon. Quickly cover the bottle opening with the glass plate. Extinguish the sulfur fire by holding the spoon under the dripping tap water. Turn off the spigot.
5. Observe the contents of the bottle for approximately three minutes. Record your observations.
6. Add 10 mL of distilled water to the smoke-filled bottle. Quickly replace the lid. Take the bottle to your laboratory bench. Swirl the contents of the bottle carefully for one minute.
7. Place a drop of *Paramecium* (or mixed) culture on a microscope slide. Examine it under the microscope. Look for the living organisms. Record your observations.
8. Add three drops of solution from the bottle to the slide. Be sure that the drops fall evenly over the original *Paramecium* drop. Observe the events happening under the microscope for three minutes. Record your observations.
9. Remove the apple skin from the bottle and note any changes in its appearance.
10. Using a stirring rod, place a drop of distilled water on pieces of red and blue litmus paper and on a piece of pH paper. Record your observations.
11. Pour about 2 mL of the liquid from the glass bottle into a test tube. Use a stirring rod to test the solution with each color of litmus paper and with pH paper. Record your observations. Acid rain has a pH of 4–4.5. Is your solution more or less acidic than this?
12. Drop a 1-cm length of magnesium ribbon into the test tube. Observe it for at least three minutes. Record your observations.
13. Add two marble chips to the solution in your bottle. Observe for at least three minutes. Record your observations.

Questions

1. Write an equation for the burning of sulfur in air.
2. When the gas produced by burning sulfur dissolved what effect did it have on the acidity of the distilled water?
3. Describe what happened to the *Paramecium* culture when the solution from the glass bottle was added. If this solution were steadily added to a small lake over a long period of time, might it affect organisms living there? Explain.
4. If liquid similar to the solution in the glass bottle were allowed to stand on a marble statue or on steel girders supporting a bridge, what effect might it have?
5. What information would you need to make you more confident of your answers to the second question in item 3 and to item 4 above?

D.13 Acids and Bases: Structure Determines Function

In 1883 the Swedish chemist Svante Arrhenius defined an acid as any substance which, when dissolved in water, generates hydrogen ion (H^+). He defined a base as a substance that generates hydroxide ion (OH^-) in water solutions. It is the hydrogen ion in solution ($H^+(aq)$) that makes a solution acidic and reactive enough to cause damage. The equations for formation of solutions of some common acids are:

$$HCl(g) \xrightarrow{water} H^+(aq) + Cl^-(aq)$$

$$HNO_3(l) \xrightarrow{water} H^+(aq) + NO_3^-(aq)$$

$$H_2SO_4(l) \xrightarrow{water} H^+(aq) + HSO_4^-(aq)$$

Most common acids are molecular compounds that contain covalently bonded hydrogen atoms. At least one hydrogen atom in each molecule is bonded in a way that the hydrogen atom can be attracted away in water solution. The hydrogen atom leaves behind an electron and is transferred to a water molecule as a hydrogen ion (H^+). As shown above, the remainder of the acid molecule becomes an anion.

Many common bases, unlike acids, are ionic compounds—they contain ions to begin with. As with all soluble ionic compounds, when such a base dissolves in water, the ions separate from each other and disperse uniformly in the solution. The hydroxide ion (OH^-) in solution imparts the properties we associate with basic solutions. Equations for dissolving some bases are as follows:

$$NaOH(s) \xrightarrow{water} Na^+(aq) + OH^-(aq)$$

$$KOH(s) \xrightarrow{water} K^+(aq) + OH^-(aq)$$

$$Ba(OH)_2(s) \xrightarrow{water} Ba^{2+}(aq) + 2\ OH^-(aq)$$

The reaction between equal amounts of $H^+(aq)$ and $OH^-(aq)$ to produce water results in disappearance of both the acidic and basic properties. This **neutralization** reaction can be represented by a total ionic equation. For the neutralization of hydrochloric acid ($HCl(aq)$) by sodium hydroxide ($NaOH(aq)$), the total ionic equation is:

$$H^+(aq) + Cl^-(aq) + Na^+(aq) + OH^-(aq) \rightarrow H_2O(l) + Cl^-(aq) + Na^+(aq)$$

Note that Na^+ and Cl^- ions appear on both sides of the equation. They are sometimes called **spectator ions** because they do not take part in the reaction. Omitting them from the equation leaves a **net ionic equation,** which shows only the reaction that takes place. The net ionic equation for this reaction is:

$$H^+(aq) + OH^-(aq) \rightarrow H_2O(l)$$

You can see why the acidic and basic properties of the solutions disappear. After a neutralization reaction, only water, a neutral substance, remains.

Some of the acids in acid rain, such as sulfurous acid (H_2SO_3), nitrous acid (HNO_2), and carbonic acid (H_2CO_3), are weak. Others, such as sulfuric acid (H_2SO_4) and nitric acid (HNO_3), are strong. A common base in the environment, ammonia (NH_3), is a weak base. The next section explains why some acids and bases are stronger than others.

D.14 Acid and Base Strength

Acids and bases are classified as weak or strong according to how readily they produce hydrogen ions ($H^+(aq)$) or hydroxide ions ($OH^-(aq)$) in water. Note that we are not referring here to the amounts of acid or base in solution, but to the nature of the individual acids or bases themselves.

The more readily an acid produces hydrogen ions in water solution, the stronger an acid it is. When a "strong acid" dissolves in water, almost every molecule splits into a hydrogen ion and an anion. Nitric acid, for example, is a strong acid; formation of a nitric acid solution is represented by the following equation:

$$HNO_3(l) \xrightarrow{water} H^+(aq) + NO_3^-(aq)$$

In a nitric acid solution there are very few HNO_3 molecules and large numbers of hydrogen and nitrate ions.

In a "weak acid" only a few acid molecules split into hydrogen ions and anions; most acid molecules remain as molecules in solution. Nitrous acid is a weak acid. The equation for formation of a nitrous acid solution is written:

$$\underset{\text{molecular form}}{HNO_2(l)} \xrightleftharpoons{water} \underset{\text{ionized form}}{H^+(aq) + NO_2^-(aq)}$$

A solution of nitrous acid always contains many more HNO_2 molecules than $H^+(aq)$ and $NO_2^-(aq)$ ions. Note the double arrow written in the equation above. One meaning of a double arrow is that the reactants are not completely consumed, but that the solution contains both reactants and products. (Most often a double arrow means the reaction is in chemical equilibrium. In a chemical reaction at equilibrium both the forward reaction and its reverse occur at the same rate. The result is unchanging concentrations of all reactants and products.)

Ionic bases, such as sodium hydroxide (NaOH) and potassium hydroxide (KOH), are strong bases—their solutions contain only cations and OH^- ions. However, the concentration of OH^- ions in solution sometimes is limited by the low solubility of the base. Magnesium hydroxide ($Mg(OH)_2$) is such a base.

A common weak base is ammonia, which forms OH^- ions in solution by the following reaction:

$$NH_3(g) + H_2O(l) \rightleftharpoons NH_4^+(aq) + OH^-(aq)$$

YOUR TURN

VI.11 Acids and Bases

1. The following strong acids are found in acid rain. For each, give the name and write an equation for the formation of ions in aqueous solution.
   a. HNO_3. The nitrate ions from its ionization can be used to fertilize plants.
   b. H_2SO_4. One of its ionization products is responsible for damaging buildings and monuments.

2. The following weak acids are found in acid rain. For each, give the name and write an equation for the formation of ions in aqueous solution.
   a. H_2CO_3. Decomposition of this acid gives the fizz to carbonated soft drinks.
   b. H_2SO_3. A component of acid rain that damages plants and man-made structures.

3. Each of the following strong bases has importance in commercial and industrial applications. Give the name of each and write an equation showing the ions formed in aqueous solution.
   a. $Mg(OH)_2$. This is one of the compounds in the mineral magnesite. It is also the active ingredient in milk of magnesia, an antacid for upset stomachs.
   b. $Al(OH)_3$. This compound is used to affix dyes to fabrics.
4. The following equation represents the formation of an aqueous solution of methylamine (CH_3NH_2), a weak organic base used in the chemical industry. Which species shown are present in an aqueous solution of methylamine?

$$CH_3NH_2(g) + H_2O(l) \rightarrow CH_3NH_3^+(aq) + OH^-(aq)$$

5. Write the total ionic equation for the reaction between $HNO_3(aq)$ and $Mg(OH)_2(s)$. This is one reaction that occurs when acid rain falls on the mineral magnesite. Then write the net ionic equation.

D.15 pH

Water and all its solutions contain both hydrogen and hydroxide ions. In pure water and neutral solutions, the concentrations of these ions are very small but equal. In acid solutions, the hydrogen ion concentration is high and the hydroxide concentration is extremely low. In basic solutions, the hydroxide ion concentration is high and the hydrogen ion concentration is extremely low. The pH concept is built on the relationship between hydrogen ions and hydroxide ions in water. The term pH stands for the "power of hydrogen ion"—a solution in which the hydrogen ion concentration is 10^{-3} M has a pH of 3. (The concept of pH was introduced in the water unit, in Part C.6.)

A pH value of 7 in a water solution at 25 °C represents a neutral solution, one where the concentrations of $H^+(aq)$ and $OH^-(aq)$ are both equal to 10^{-7} M. The pH of pure water is 7.

Values of pH lower than 7 represent acidic solutions. The lower the pH value, the greater the acidity of the solution. In an acidic solution, the concentration of H^+ ion is greater than that of OH^- ion. An acidic solution with a pH of 1 has a hydrogen ion concentration of 10^{-1} M. This is more acidic than a solution of pH 2, where the hydrogen ion concentration is 10^{-2} M.

In contrast, values above pH 7 represent basic solutions. A solution with a pH of 14 has the very low H^+ concentration of 10^{-14} M and an OH^- concentration of 1 M. It is considerably more basic than a solution of pH 8, which has an H^+ concentration of 10^{-8} M and an OH^- concentration of 10^{-6} M.

As you see, with each step up or down the pH scale the acidity changes by a factor of 10. Thus, lemon juice, pH 2, is 10 times more acidic than a soft drink, at pH 3, which is 10,000 times more acidic than pure water. (Figure I.24 in the water unit gives typical pH values for some common materials.)

YOUR TURN

VI.12 pH

1. Following are listed some common aqueous solutions with their typical pH values. Classify each as either acidic, basic, or neutral. Arrange them in order of increasing hydrogen ion concentration.
    a. Stomach fluid, 1
    b. A solution of baking soda, 9
    c. A cola drink, 3
    d. A solution of household lye, 13
    e. Milk, drinking water, 6
    f. Sugar dissolved in pure water, 7
    g. Household ammonia, 11

2. How many more times acidic is a cola drink than milk?

PART D
SUMMARY QUESTIONS

1. Identify the major types of air pollutants.
2. In what sense is "pollution free" combustion an impossibility?
3. Identify the major components of smog and their sources.
4. Define the term "synergism" and explain its relevance to air pollution.
5. In which region of the United States is acid rain most prevalent? Why?
6. Name the major substances responsible for acid rain. What are their sources?
7. How might efforts to control air pollution result in other kinds of pollution?
8. Pollution has sometimes been defined as "a resource out of place." Name a substance that is a resource in one part of the atmosphere and a pollutant when found in another part.
9. Write the ionic equation for the reaction of NO_2 with rainwater.
10. Which of the following compounds do you recognize as acids? as bases?
    a. NaOH
    b. HNO_3
    c. CH_4
    d. $C_{12}H_{22}O_{11}$ (table sugar)
    e. H_2SO_3
11. Which of the following solutions has the lowest pH? the highest pH?
    a. lemon juice
    b. stomach fluid
    c. drain cleaner (NaOH)
12. Technology to prevent acid rain is available. Why, then, is acid rain still a problem?

EXTENDING YOUR KNOWLEDGE (OPTIONAL)

- Carbon monoxide can interfere with the body's O_2 transport and exchange system. Investigate the health effects of CO and its relationship to traffic accidents.
- Investigate the advantages and disadvantages of various alternatives to the standard internal combustion engine. Options include the electric engine, gas turbine, Rankine engine, stratified charge engines, Wankel engine, Stirling engine, diesel engine, and expanded mass transportation systems.
- Make an acid-base indicator at home with red cabbage juice. (Consult your teacher for instructions.)
- Find out how acidic rainwater is in your locality. Wash a plastic (not glass) container with 6 M HCl, and rinse it thoroughly with deionized water. Test the rinse water with pH paper to be sure all the HCl has been rinsed out. Then collect some rainwater. Filter the rainwater through clean filter paper and determine its acidity with pH paper or with a pH meter. If you cannot analyze it immediately, refrigerate it in a sealed container to prevent a change in pH. (What might cause such a change?)

An environmental scientist at the Department of Energy's Argonne National Laboratory uses sophisticated instrumentation to measure the effect of legally permissable levels of atmospheric sulfur dioxide on soybean growth and yield.

E PUTTING IT ALL TOGETHER: IS AIR A FREE RESOURCE?

Thus far we have investigated the general properties of gases, the structure and functions of the atmosphere, and how human activity may alter the atmosphere. We have also surveyed ways to control air pollution. However, we have not considered the overall success of such efforts. In fact, how clean should the air be? At what cost? These concerns will also be considered in this final activity.

E.1 Air Pollution Control: A Chemical Success Story?

In Section D.3 you were introduced to the main culprits of air pollution. You examined the relative amounts of total suspended particles (TSP), sulfur oxides (SO_x), nitrogen oxides (NO_x), hydrocarbons (HC), and carbon monoxide (CO) emitted by various sectors of our economy. Later several ways of controlling these emissions were described. Most of these control technologies were instituted since the early 1970s. How successful have they been?

Figure VI.18 presents information summarizing U.S. pollutant additions to the atmosphere from 1975–1983, estimated by the Environmental Protection Agency. Refer to the figures to answer these questions.

1. Estimate the overall percent change in total emissions for each of the four pollutants using the following formula:

$$\text{Percent change} = \frac{1983 \text{ value} - 1975 \text{ value}}{1975 \text{ value}} \times 100$$

   a. Do the overall trends represent an improvement or a worsening of air quality since 1975?
   b. From 1975 to 1983 an increase in economic activity led to increased demand for energy and products. For example, during this period the distance traveled by U.S. vehicles increased about 20%. Why is it difficult to improve air quality when energy consumption is increasing?

2. On a separate sheet of paper, complete a table similar to the one below. Compare 1975 emissions data with those from 1983 by indicating how specific sources contributed to the overall emissions of each pollutant. The first category has been completed for you.

Data table: Shift in emission patterns, 1975 to 1983			
Pollutant	**Increased**	**Unchanged**	**Decreased**
SO_x	Transportation	None	Fuel combustion; industrial processes
TSP			
NO_x			
HC			

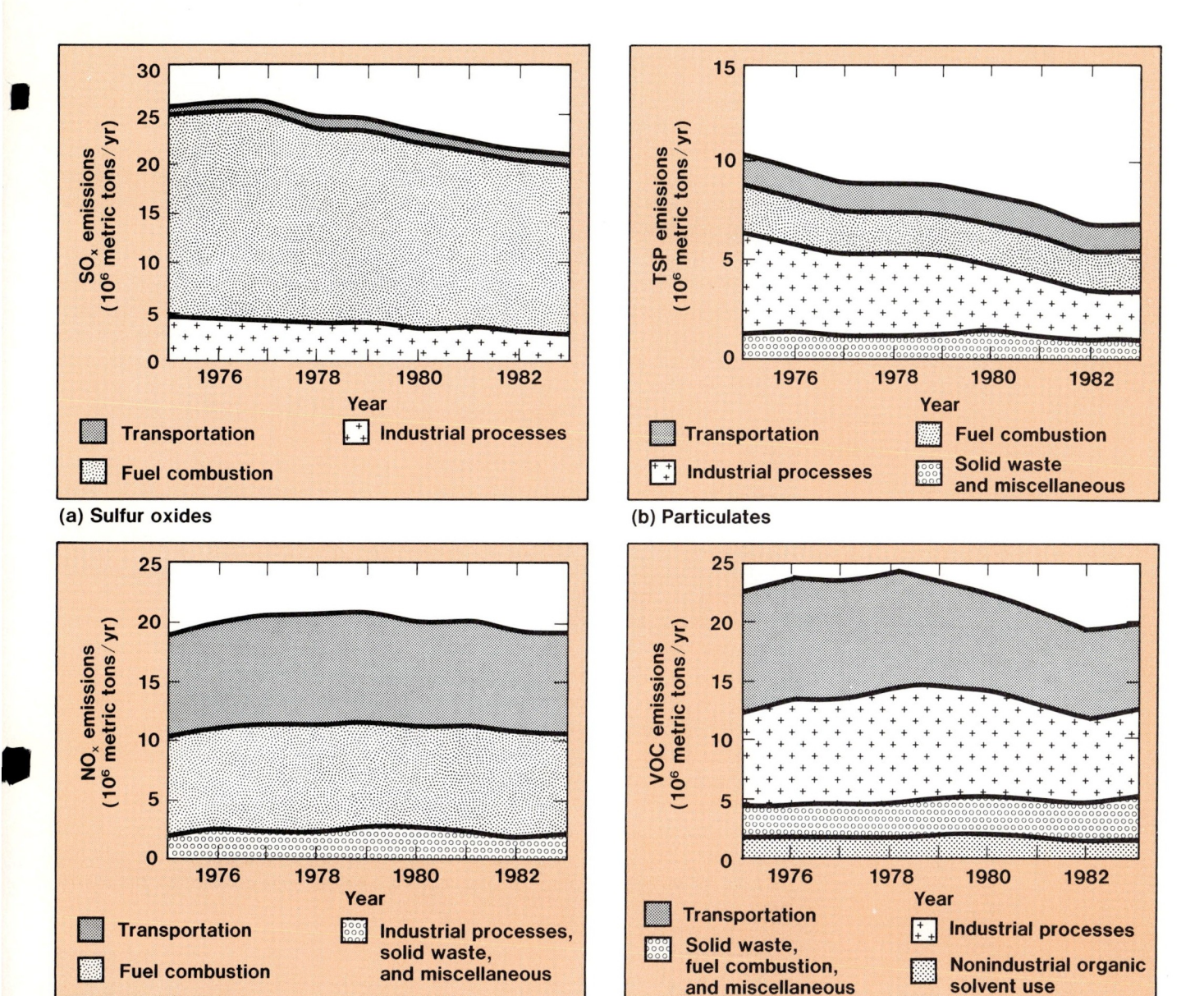

Figure VI.18 National trends in the emission of four air pollutants from 1975–1983; (a) sulfur oxides (SO$_x$); (b) particulates (TSP, total suspended particulates); (c) nitrogen oxides (NO$_x$); (d) volatile organic compounds (VOC; primarily hydrocarbons).

U.S. Environmental Protection Agency, Washington, 1985

a. What factors could account for a decrease in the total emissions of each pollutant from 1975 to 1983?
b. What factors could account for an increase in emissions of a given pollutant from any particular source?
c. Note that some sources emitted the same quantity of a given pollutant in 1983 as in 1975. Does this necessarily indicate that no new pollution control methods were used? Why or why not?

3. Do you agree or disagree with the following statement? "Great strides have been made in improving the quality of the air we breathe." Explain your answer. In addition to emissions data, what evidence would be useful in judging the effects of pollution control technologies?

4. What information would you want to consider before you decided to agree or disagree with the following statement? "Stricter laws and better methods are needed to control air pollution."

E.2 Paying the Price

Earlier in this unit we noted that air pollution costs the United States an estimated $16 billion per year. This financial loss shows up in destroyed crops, weakened livestock, corrosion of buildings and machines, and higher workmen's compensation and hospital costs for those with air-pollution-related diseases. However, this does *not* include less visible, but potentially disastrous long-term costs associated with altering nature's cycles.

Of course, pollution control has its own costs. Whether one considers catalytic converters in automobiles or scrubbers in power plants, pollution control technologies increase the costs of material goods and energy. In other words, using the atmosphere as a "free" resource and repository actually involves real expense. With or without pollution control, citizens in industrialized nations pay a price for the quality of air they breathe.

Economists deal with pollution control policy issues (as they do many issues) in terms of cost-benefit analyses. Each consequence of a given policy is reduced to a common denominator: dollars. The total cost of any air pollution control policy is the sum of two elements. One is **damage costs**: tangible losses, such as those discussed above, and intangible costs, such as reduced visibility and respiratory irritation. The other element consists of **control costs**—the direct costs of pollution control methods along with any indirect costs, such as unemployment created by a plant shutdown or relocation.

The overall result of the costs and benefits of pollution control are combined in plots like those in Figure VI.19. The plots start at the left with pollution levels reduced practically to zero (high air quality). As you might expect, at this point the cost of the control measures is the highest. Also, at the highest air quality on the left, the cost of pollution damage is very low. Note the changes as air quality is allowed decrease, represented by moving from left to right on the figure. Control costs decrease as the level of pollution is allowed to increase. However, simultaneously, damage costs increase with increasing pollution levels.

The top curve represents the total cost to society of pollution controls and pollution damage. Look at the vertical line marked on the plot. At the pollution level shown by point *a*, the damage costs are given by the distance *ab*, the control

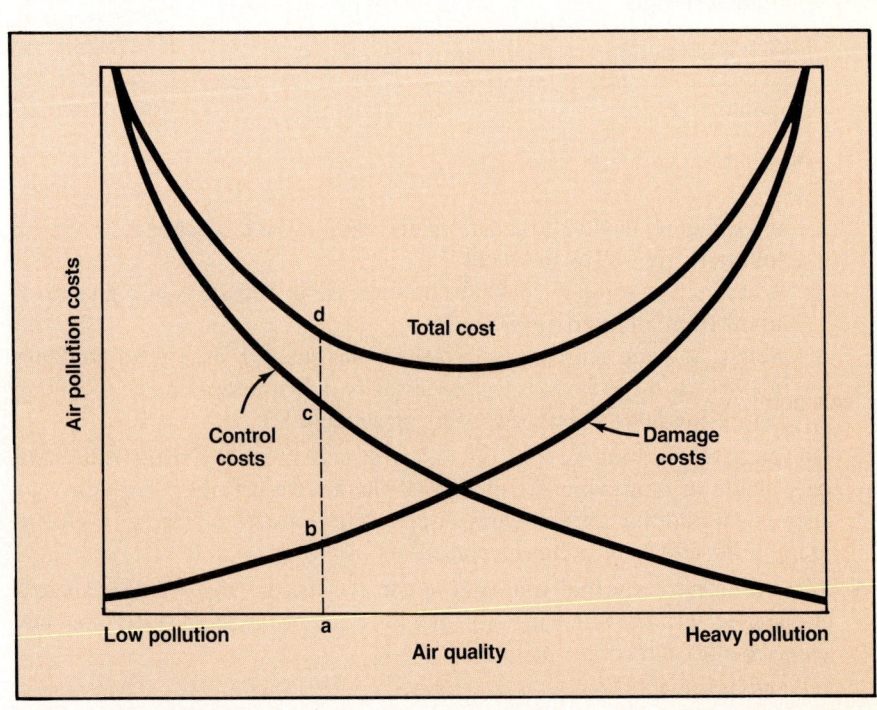

Figure VI.19 The cost of air pollution.

Source: Seinfield, John H., *Air Pollution: Physical and Chemical Fundamentals*, p. 41. Copyright ©1975 by McGraw-Hill, Inc. Reprinted with permission.

costs are given by the distance *ac,* and the total costs to society are given by *ad,* equal to the sum of *ab* plus *ac.* At this point the cost to society for control is greater than the cost due to damage.

Theoretically, the best balance of control costs and damage costs is where the total cost curve reaches a minimum. This point, will, of course, be different for different pollution and control combinations. Such a cost-benefit analysis is most straightforward for a single business that will bear both the costs of control and damage. Often pollution control analyses are more complicated because the costs of control and damage do not apply to the same group.

Based on what you have learned in this unit and the relationships illustrated in Figure VI.19, answer the following questions:

1. At low pollution levels (high air quality) which costs are highest? Which are lowest? Explain your answers.
2. At high pollution levels (low air quality), which costs are highest? Which are lowest? Explain your answers.
3. Is it possible, technologically or economically, to eliminate all pollution or damage costs? Why?
4. How would the development of new control technologies (such as catalytic converters which use cheaper, more readily available metals) affect the control costs curve?
5. Which costs (damage costs or control costs) do you think can be estimated with greater certainty? Why? What are some difficulties in estimating the other kind of cost?
6. For each point on the graph, the total cost equals the sum of control costs and damage costs. Many economists would say the objective of pollution control is to minimize the total cost. What assumptions are built into such a cost-benefit analysis? Do you think these assumptions are valid? Why?
7. Counting only direct costs that can be measured easily in dollars, the damage costs of U.S. air pollution are currently calculated as $68.38 per year for each individual. Do you think this figure is a fair total cost figure? Why?
8. Automobile emissions of CO, HC, and NO_x have been reduced (respectively) by 96%, 95%, and 71% compared to pre-1968, uncontrolled levels. Assuming you favor stricter pollution control, how much would you be willing to pay (on the price of a new car or for modifying an old one) to decrease the NO_x emissions an additional 25%? When considered on a personal level, control costs seem much higher than damage costs. Why?
9. Are the benefits of cleaner air shared equally by all? Are the control costs of cleaner air shared equally? Why?
10. How much (in dollars) is it worth to you to see a clear blue sky? How much is it worth to you to smell fresh, clean air? Are these questions fair? Are they important?

E.3 Looking Back and Looking Ahead

At this point, you have explored the chemistry of matter and energy interactions in the Earth's atmosphere, land, and water. You have also considered some social issues surrounding our use of these portions of our world as resources and repositories.

The previous units also have been concerned with a fourth "sphere"—the **biosphere**—supported by vital interactions with the other three. In the next unit you will explore the chemistry within the marvelously intricate "ecosystem" of the human body. You will see how individual and social decisions can affect our well-being. As you might suspect, chemistry continues to help us make wiser decisions and define questions for further research. You'll enter the world of "personal chemistry" in the next unit.

FMC Corporation: Air Quality Control Operation

The crane is setting the seven-ton top section on a 105-foot-tall absorber—part of the air pollution control system for Northern Indiana Public Service Company's Rollin M. Schahfer Station near Wheatfield, Indiana. Each absorber uses a sodium-based scrubbing solution to absorb up to 3 tons of sulfur dioxide per hour.